T0329559

Viscoplastic Flow in Solids Produced by Shear Banding

Viscoplastic Flow in Solids Produced by Shear Banding

Ryszard B. Pęcherski

 Institute of Fundamental Technological Research
Polish Academy of Sciences, Warsaw, Poland

Registered Office(s)
John Wiley & Sons, Inc., 111 River Street, Hoboken, NJ 07030, USA
John Wiley & Sons Ltd, The Atrium, Southern Gate, Chichester, West Sussex, PO19 8SQ, UK

Editorial Office
111 River Street, Hoboken, NJ 07030, USA

For details of our global editorial offices, customer services, and more information about Wiley products visit us at www.wiley.com.

Wiley also publishes its books in a variety of electronic formats and by print-on-demand. Some content that appears in standard print versions of this book may not be available in other formats.

Library of Congress Cataloging-in-Publication Data

Names: Pęcherski, Ryszard B., author. | John Wiley & Sons, publisher.
Title: Viscoplastic flow in solids produced by shear banding / Ryszard B. Pęcherski.
Description: Hoboken, NJ : Wiley, 2022. | Includes bibliographical references and index.
Identifiers: LCCN 2022010729 (print) | LCCN 2022010730 (ebook) | ISBN 9781119618584 (cloth) | ISBN 9781119618607 (adobe pdf) | ISBN 9781119618638 (epub)
Subjects: LCSH: Shear (Mechanics). | Deformations (Mechanics). | Viscoplasticity.
Classification: LCC TA417.7.S5 P44 2022 (print) | LCC TA417.7.S5 (ebook) | DDC 620.1/1245–dc23/eng/20220408
LC record available at https://lccn.loc.gov/2022010729
LC ebook record available at https://lccn.loc.gov/2022010730

Cover Design: Wiley
Cover Image: © Sergey Ryzhov/Shutterstock

Set in 9.5/12.5pt STIXTwoText by Straive, Pondicherry, India
Printed and bound by CPI Group (UK) Ltd, Croydon, CR0 4YY

C9781119618584_170522

Contents

Preface

The thorough Investigations of the new types of materials – nano- and ultrafine-grained metallic solids, amorphous metal alloys called glassy metals, and high-performance alloys – lead to an essential general conclusion. Observing their failure processes, one may notice that a paradigm shift transpires before our eyes regarding the widely known and accepted ductile failure micromechanisms as *initiation, growth, and coalescence of voids*. The recent nonstandard experiments confirm the novel observations about the vital importance of accompanying shear modes, e.g. stereo digital image correlation, the tomograms of X-ray, or synchrotron techniques related to 3D imaging methods. Dunand and Mohr (2010), using two- and three-dimensional digital image correlation technique, measured the surface strain fields on tensile specimens, including the one with a central hole and circular notches. The samples came from TRIP780 steel sheets. The authors concluded that the non-porous plasticity model's numerical predictions agree well with all macroscopic measurements for various loading conditions. Dunand and Mohr (2011) studied for TRIP780 steel the shear-modified Gurson model's predictive capabilities (Nielsen and Tvergaard 2010) and the modified Mohr–Coulomb fracture model (Bai and Wierzbicki 2008). The result is that significant differences between the two models appear with the less accurate prediction for the shear-modified Gurson model. Gorij and Mohr (2017) present a new micro-tension and micro-shear testing technique applying aluminium alloy 6016-T4 flat dogbone-shaped, as well as notched and central hole samples and smiley-shear micro-specimens to identify the parameters of hardening law and fracture initiation model. The Hosford–Coulomb damage indicator model predicts the ductile fracture initiation that appears imminent with the onset of shear localisation.

It became then evident that the known porous material models, e.g. by Shima et al. (1973), Shima and Oyane (1976), or Gurson (1977) extended by Tvergaard and Needleman (1984), reveal limited applications besides the cases when high triaxiality states are prevalent. Therefore, the studies of inelastic deformation and failure of materials should require, in my view, a fresh and novel approach. It aims towards a better understanding and description of the multilevel character of shear deformation modes. It is also worth stressing that Pardoen (2006) emphasizes the role of shear localisation in low-stress triaxiality ductile fracture.

The known experimental data reveal that metallic solids' inelastic deformation appears in the effect of competing mechanisms of slips, twinning, and micro-shear banding. Shear banding is a form of instability that localises large shear strains in relatively thin bands. The micro-shear bands transpire as concentrated shear zones in the form of transcrystalline layers of the order 0.1 µm thickness. The observations show that a particular micro-shear band operates only once and develops rapidly to its full extent. The micro-shear bands, once formed, do not contribute further to the increase of inelastic shear strain. Thus, it appears that successive generations of active micro-shear bands, competing with the mechanisms of multiple crystallographic slips or twinning, are responsible for the inelastic deformation of metals. Therefore, identifying the physical origins of the initiation, growth, and evolution of micro-shear bands is fundamental for understanding polycrystalline metallic solids' macroscopic behaviour.

A new physical model of multilevel hierarchy and evolution of micro-shear bands is at the centre of this work. An original idea of extending the representative volume element (RVE) concept using the general theory of propagation of the singular surfaces of microscopic velocity field sweeping the RVE appears useful for the macroscopic description of shear-banding mechanism in viscoplastic flow, cf. Pęcherski (1997, 1998). The essential novelty of the presented approach comes from numerous observations revealing that the process of shear banding is **the driving factor – a cause and not a result.** So it turns out, in my view, that the successive generations of micro-shearing processes induced mostly by changing deformation path produces and controls viscoplastic flow. On the other hand, one may recall many valuable papers containing the results of in-depth analysis, modelling of dislocation-mediated multi-slip plastic deformation, and numerical simulations of the laminate microstructure, bands, or shear strain localisation in crystalline solids cf. Dequiedt (2018), Anand and Kothari (1996), Havner (1992), as well as Petryk and Kursa (2013) and the wealth of papers cited herein.

Recent studies reveal that two types of shear banding, generating the inelastic deformation in materials, can play a pivotal role.

- The first type corresponds to *the rapid formation of the multiscale shearbanding systems*. It contains micro-shear bands of the thickness of the order of the $0.1\,\mu$m, which form clusters. The clusters propagate and produce the discontinuity of microscopic velocity field v_m. They spread over the RVE of a traditional polycrystalline metallic solid. A detailed discussion of such a case is presented in Pęcherski (1997, 1998). A new concept of the RVE with a strong singularity appears, and the *instantaneous shearbanding contribution function f_{SB}* originates.

- The second type is *a gradual, cumulative shear banding* that collects microshear bands' particular contributions and clusters. Finally, they accumulate in the localisation zone spreading across the macroscopic volume of considered material. Such a deformation mechanism appears in amorphous solids as glassy metals or polymers. It seems that there are the local shear transformation zones (STZs) behind the cumulative kind of shear banding, cf. Argon (1979, 1999), Scudino et al. (2011), and Greer et al. (2013). The volumetric contribution function f_{SB}^{v} of shear banding appears in such a case.

Often both types of the above-mentioned shearing phenomena appear with variable contribution during the deformation processes. During shaping operations, this situation can arise in polycrystalline metallic solids, typically accompanied by a distinct change of deformation or loading paths or a loading scheme. Also, materials revealing the composed, hybrid structure characterizing with amorphous, ultra-fine grained (ufg), and nanostructural phases are prone to the mixed type of shear banding responsible for inelastic deformation, cf. the recent results of Orava et al. (2021) and Ziabicki et al. (2016).

The commonly used averaging procedures over the RVE need deeper analysis to account for the multilevel shear-banding phenomena. The RVE of crystalline material is the configuration of a body element idealized as a particle. The particle becomes a carrier of the inter-scale shearing effect producing the viscoplastic flow. It leads to an original and novel concept of the particle endowed with the transfer of information on a multilevel hierarchy of micro-shear bands developing in the body element of crystalline material. The discussion about the difficulties and shortcomings of applying a traditional direct multiscale integration scheme appears in Chapter 4. The remarks mentioned above motivate the core subject of the work and underline the new way of thinking.

Ryszard B. Pęcherski
2022
Kraków and Warszawa, Poland

Acknowledgements

I want to express my gratitude for the helpful and friendly guidance offered during my writing efforts shown by Wiley's competent and patient staff led by Ms Juliet Booker, Managing Editor. Thank you very much for accompanying me on my long journey to navigate the bumpy roads of British syntax and phraseology. In such a case, the role of my cicerone – Ms Nandhini Tamilvanan, Content Refinement Specialist – appeared invaluable. Last but not least, acknowledgement belongs to the Creative Services Team coordinated by Ms Becky Cowan, Editorial Assistant, in preparing the book's cover. Their professionalism led me to choose the motif that sheds new light in Chapter 1 on the relevant issues related to industrial applications.

References

Anand, L. and Kothari, M. (1996). A computational procedure for rate-independent crystal plasticity. *J. Mech. Phys.* Solids. 44: 525–558.

Argon, A.S. (1979). Plastic deformation in metallic glasses. *Acta Metall.* 27: 47–58.

Argon, A.S. (1999). Rate processes in plastic deformation of crystalline and noncrystalline solids. In: *Mechanics and Materials: Fundamentals and* (ed. M.A. Linkages, R.W.A. Meyers and H. Kirchner), 175–230. New York: Wiley.

Bai, Y. and Wierzbicki, T. (2008). A new model of metal plasticity and fracture with pressure and Lode dependence. *Int. J. Plast.* 24: 1071–1096.

Dequiedt, J.L. (2018). The incidence of slip system interactions on the deformation of FCC single crystals: system selection and segregation for local and non-local constitutive behavior. *Int. J. Solids Struct.* 141–142: 1–14.

Dunand, M. and Mohr, D. (2010). Hybrid experimental–numerical analysis of basic ductile fracture experiments for sheet metals. *Int. J. Solids Struct.* 47: 1130–1143.

Dunand, M. and Mohr, D. (2011). On the predictive capabilities of the shear modified Gurson and the modified Mohr–Coulomb fracture models over a wide range of stress triaxialities and Lode angles. *J. Mech. Phys.* Solids. 59: 1374–1394.

Gorij, M.B. and Mohr, D. (2017). Micro-tension and micro-shear experiments to characterize stress-state dependent ductile fracture. *Acta Mater.* 131: 65–76.

Greer, A.L., Cheng, Y.Q., and Ma, E. (2013). Shear bands in metallic glasses. *Mater. Sci. Eng., R.* **74**: 71–132.

Gurson, A.L. (1977). Continuum theory of ductile rupture by void nucleation and growth. I. Yield criteria and flow rules for porous ductile media. *J. Eng. Mater. Technol.* 99: 2–15.

Havner, K.S. (1992). *Finite Plastic Deformation of Crystalline Solids.* Cambridge University Press.

Nielsen, K.L. and Tvergaard, V. (2010). Ductile shear failure of plug failure of spot welds modeled by modified Gurson model. *Eng. Fract. Mech.* 77: 1031–1047.

Orava, J., Balachandran, S., Han, X. et al. (2021). In situ correlation between metastable phase-transformation mechanism and kinetics in a metallic glass. *Nat. Commun.* 12: 2839. https://doi.org/10.1038/s41467=021-23028-9.

Pardoen, T. (2006). Numerical simulation of low stress triaxiality of ductile fracture. *Comput. Struct.* 84: 1641–1650.

Pęcherski, R.B. (1997). Macroscopic measure of the rate of deformation produced by micro-shear banding. *Arch. Mech.* 49: 385–401.

Pęcherski, R.B. (1998). Macroscopic effects of micro-shear banding in plasticity of metals. *Acta Mech.* 131: 203–224.

Petryk, H. and Kursa, M. (2013). The energy criterion for deformation banding in ductile single crystals. *J. Mech. Phys. Solids.* 61: 1854–1875.

Scudino, S., Jerliu, B., Pauly, S. et al. (2011). Ductile bulk metallic glasses produced through designed heterogeneities. *Scr. Mater.* 65: 815–818.

Shima, S. and Oyane, M. (1976). Plasticity for porous solids. *Int. J. Mech. Sci.* 18: 285–291.

Shima, S., Oyane, M., and Kono, Y. (1973). Theory of plasticity for porous metals. *Bull. JSME.* 16: 1254–1262.

Tvergaard, V. and Needleman, A. (1984). Analysis of the cup-cone fracture in a round tensile bar. *Acta Metall.* 32: 157–169.

Ziabicki, A., Misztal-Faraj, B., and Jarecki, L. (2016). Kinetic model of non-isothermal crystal nucleation with transient and athermal effects. *J. Mater. Sci. 51*: 8935–8952.

1

Introduction

1.1 The Objective of the Work

The subject of the book evolved since the 1990s from the many years' studies, in several joint research projects conducted together with the investigation group of Andrzej Korbel and Włodzimierz Bochniak, professors at the Faculty of Non-Ferrous Metals of the AGH University of Science and Technology in Kraków, Poland (formerly Akademia Górniczo – Hutnicza, in English: Academy of Mining and Metallurgy), cf. Figure 1.1. It concerned physics and theoretical description of deformation processes in metals, particularly in hard deformable alloys. The long-time joint efforts to understand the physical mechanisms responsible for observed phenomena coined the subject of this work. Many years of investigations of metal-forming processes based on multilevel observations – on a macroscopic scale with the naked eye, microscopic ones using optical microscopy, high-resolution transmission electron microscopy, and scanning electron microscopy – led to the critical conclusion. The traditional approach of classical plasticity theory based solely on crystallographic slip and twinning in separate grains is inadequate for predicting and modelling observed deformation processes. Such an observation played a pivotal role in developing an innovative metal-forming method called KOBO, the acronym of inventors names 'Korbel' and 'Bochniak'. This book attempts to provide theoretical foundations and empirical evidence of viscoplastic flow produced by shear banding. In the future, the presented results should make the basis for the formulation of computer codes necessary for numerical simulations of deformation processes in industrial applications. It seems that this book might fill at least partly the mentioned gap.

Viscoplastic Flow in Solids Produced by Shear Banding, First Edition. Ryszard B. Pęcherski.
© 2022 John Wiley & Sons Ltd. Published 2022 by John Wiley & Sons Ltd.

Figure 1.1 The historical AGH UST emblem. *Source:* AGH University of Science and Technology (https://www.agh.edu.pl/en/university/history-and-traditions/emblem-and-symbols/).

1.2 For Whom Is This Work Intended?

The book's readers may be graduate and postgraduate students in engineering, particularly material science and mechanical engineering. Researchers working on the physical foundations of inelastic deformation of metallic solids and numerical simulations of manufacturing processes could also benefit from this study. The content of the work is also directed at specialists in the field of rational mechanics of materials. The prerequisite knowledge of material science and continuum mechanics with related mathematical foundations, as vector and tensor algebra and tensor analysis, will appear helpful for the readers. The fundamental background may provide the recent work written by eminent scholars of great experience, Morton E. Gurtin, Eliot Fried, and Lallit Anand (Gurtin et al. 2009). Also, a modern and integrated study across the different observation scales of the foundation of solid mechanics applied to the mathematical description of material behaviour presented in the pivotal work (Asaro and Lubarda 2006) is recommendable for the readers. These works comprehensively cover the subject of rational thermomechanics, being the contemporary approach of classical treatises 'standing on the shoulders of giants' (https://en.wikipedia.org/wiki/Standing_on_the_shoulders_of_giants), cf. Chapter 4 for the discussion of a historical thread.

1.3   State of the Art

1.3.1   Motivation Resulting from Industrial Applications

Korbel and Szyndler (2010) presented an overview of the Polish engineering inventions' contribution to metal-forming technologies. Three industrial sectors can play an important role: electrical power plants, transportation, and natural environment protection. First of all, one should focus on high-quality and energy-saving extrusion and forging processes of the elements made of structural steel, non-ferrous metals, and light alloys used to produce parts of machines and other equipment manufactured by all industry sectors.

There is a need and necessity to implement innovative technical and technological solutions into metal-forming practice, making production more efficient, energy-saving, and less expensive. So, we face three challenges with new, non-conventional technologies, such as metal-processing technology in the cyclically variable plastic deformation – known as the KOBO method, cf. the US and European patents description Korbel and Bochniak (1998). The technological solution of metal forming, the KOBO method, satisfies both demands: low manufacturing costs and control of the metal substructure properties in a single operation. The premises, at the background of the method, result from the thorough experimental studies of plastic deformation mechanisms in the course of strain path change conditions (Korbel and Szyndler 2010). The change in the mode of plastic flow from the crystallographic slip of dislocations within separate grains into trans-granular localised shear (shear banding) and associated decrease of metal hardening play a controlling role in the KOBO method. Figure 1.2 illustrates the extrusion process controlled by strain path change due to the reversible twisting of the die in an oscillatory manner. The die oscillations' angle and frequency are the controlling factors of the extrusion process influencing the metal structure and mechanical properties.

Figure 1.3 shows that the load of the order of 1MN is sufficient to cold–extrusion of hardly deformable aluminium alloy 7075 into the billet form with 700 times cross-section reduction.

Due to simultaneous measurements of the extrusion force and the die-twisting torque, it was possible to evaluate the forming process's power consumption and the dependence upon the extrusion rate. The discussion on the power consumption presented in Korbel and Szyndler (2010) illustrates the method's high potential in diminishing the process's plastic work with simultaneous increase of its efficiency. To assess the global effect of

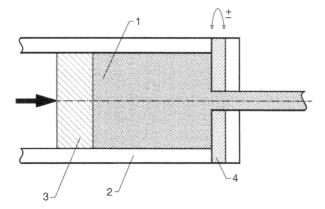

Figure 1.2 Scheme of metal extrusion throughout the oscillating die (KOBO method): 1 – billet, 2 – container, 3 – punch, 4 – oscillating die (Korbel and Szyndler 2010). *Source:* Copyright by Aleksandra Manecka – Padaż.

Figure 1.3 The pattern of the aluminium rest and extruded wire. The extrusion ratio equals 700. *Source:* Korbel and Szyndler 2010. Copyright of Włodzimierz Bochniak.

energy saving on the KOBO process, one should observe that there is no need to heat the billet higher than that in conventional metal extrusion processes. The studies of mechanical properties of extruded metals reveal additional essential features of KOBO products. Worthwhile mentioning is the unexpected thermal stability of the mechanical properties, e.g. plastic flow limit and ultimate tensile strength are not affected by heating in the temperature range where recovery processes are used to produce softening. Furthermore, hardly deformable aluminium alloys (e.g. Al 7075) and magnesium alloys (AZ31, AZ91) subjected to KOBO extrusion become superplastic at elevated temperature, cf. (Korbel and Szyndler 2010). The careful control of the KOBO-forming processes leads to the unique possibility of obtaining the extruded or forging products of the desired shape and properties. Experiments on extrusion of hardly deformable metallic materials reveal practically no limits in getting the desired shape of extrudates under 'cold deformation' conditions. Some chosen examples are displayed in Figures 1.4 and 1.5.

The paper (Bochniak et al. 2006) deals with the KOBO method of forming bevel gears from structural steel. The study's subject is a single operation of complex forging on a press with the reversible rotating die displayed in Figure 1.6. Comparing the KOBO method's forging process with the conventional ones reveals that the punch pressure and temperature are considerably lower. Despite such a reduction, the products represent the die shape correctly, the structure becomes homogeneous, and the material has desired mechanical properties (see Figure 1.4) containing an example of the regular bevel gear obtained by the KOBO method from structural steel at the studied temperature of 850 °C (Bochniak et al. 2006). Let us also recall a nice

Figure 1.4 Examples of the KOBO extrusion and forging products received in semi-industrial conditions. *Source:* Korbel and Szyndler 2010. Copyright of Włodzimierz Bochniak.

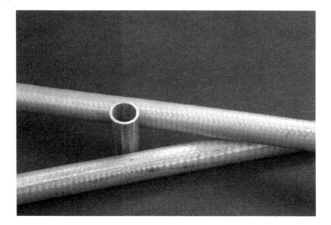

Figure 1.5 Fine tube of magnesium alloy AZ91 extruded at room temperature using 1MN load capacity press. *Source:* Korbel and Szyndler 2010. Copyright of Włodzimierz Bochniak.

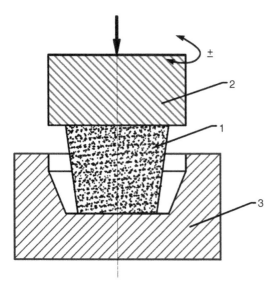

Figure 1.6 Schematic presentation of the forging process by the KOBO method: (1) forged material, (2) cyclically rotating punch, and (3) die (anvil). *Source:* (Bochniak et al. 2006). Copyright by Aleksandra Manecka – Padaż.

illustrative picture of the bevel gear displayed on the book's cover. The image provided kindly by the editorial staff comes from other sources, and the shown example of bevel gear is the traditional milling effect.

Summing up, the authors observe the KOBO method's extrusion or forging process from the structural point of view. The slips' organisation with increasing strain leads to transgranular, localised plastic strain. Such a localisation appears as growing clusters of micro-shear bands and is related to strain softening of metal. The rapid change of the loading scheme and corresponding change of the deformation path leads to instantaneous localisation of plastic flow in the shear bands irrespective of the deformation process's advancement. According to Korbel and Bochniak (2003), the mentioned procedure does not guarantee to keep this state for a long time, and the cyclic repetition of additional external agents in this process is required.

1.3.2 KOBO Processes Resulting in Viscous Effects

The experimental investigations and microscopic analysis of the substructure of metals and alloys carried out by Korbel and Bochniak investigation group led to the novel observation of viscous effects of deformation processes by the KOBO method. The results presented in the papers by Korbel et al. (2011), Bochniak et al. (2011, 2013) indicate that the point defects of supra-equilibrium concentration, generated in periodically variable conditions of plastic flow in the course of the KOBO process, play a decisive role. The massive production of point defects leads to the superplastic behaviour of metallic solid not observed in other plastic-forming methods. The authors state that: 'It seems reasonable, therefore, to conclude that die oscillation frequency (torsion of material) is the determinant of the amount of point defects and its increase should enhance the process. The occurrence of diffusing atoms or vacancies stream equalising the concentration leads to a significant decrease in viscosity of the material, generating an alternative to the dislocation slip mechanism of plastic deformation', cf. (Korbel et al. 2011), p. 2893. The analysis justifies the author's view that the mechanism of metal extrusion using the mentioned KOBO technology is induced by the intensive generation of point defects. Thus, the authors hypothesise that a viscous flow with 'Newtonian fluid' features is a dominant deformation mechanism in KOBO processes. Generally, they identify the description of deformation occurring, e.g. during extrusion by the KOBO method as viscoplastic flow. However, on the other hand, in deformable solids' mechanics, the early viscoplasticity model belongs to Bingham (1916). It shows the linear dependency of shear stress on shear strain rate:

$$\tau = \tau_0 + \mu\dot{\gamma},$$

where τ_0 is yield stress in shear and $\dot{\gamma}$ denotes the shear strain rate. Neglecting $\tau_0 = 0$, one arrives at the linear model of 'Newtonian fluid'. An analogy with magnetorheological materials appears here. From the papers of Frąś (2015),

Frąś and Pęcherski (2018), it seems that the linear Bingham model does not conform to experimental data contrary to the original nonlinear viscoplasticity model of Perzyna (1963). The similar conclusion leads the above discussion on a viscous flow resulting from the massive production of point defects activated in KOBO processes. In my view, a more comprehensive theory of the viscoplastic flow produced by shear banding is in order. The observation about the importance of viscous effects accompanying rate-dependent plastic flow during KOBO processes is accounted for in the book.

1.4 Summary of the Work Content

The preface introduces novel concepts and the framework of the book. Chapter 1 presents the motivation and leading thread of the work related to a detailed discussion of the physical basis developed in Chapter 2. This chapter contains the synthetic approach to observations that appear helpful in formulating the viscoplastic flow description in metallic solids produced by shear banding. These views are underlined in the text as the set of statements denoted Observations 2.1, 2.2, . . . 6.1, including the results of own inquiries. The heuristic foundations of the theoretical description of large inelastic deformations create the rational formulation of a multiscale system of shear bands formation. Chapter 3, on the other hand, accounts for shear banding in the continuum model of inelastic deformations. This chapter contains the results of the earlier author's investigations related to micromechanical foundations of finite plastic deformations theory accounting for the shear-banding mechanism summarised in Observation 3.1 and Hypothesis 3.1, extending the generally accepted concept of representative volume element (RVE). The extension provides the possibility of the existence in RVE of the singular discontinuity surface of order one of the microscopic velocity field on which the tangential component of velocity experiences a jump travelling at the speed V_s. Further, Chapter 4 presents the basics of rational mechanics of materials. A small historical account of rational mechanics is given here. The continuum mechanics description of shear banding is the subject of Chapter 5. The theoretical foundations of the deformation of a body due to shear banding are presented in Chapter 6. In Chapter 7, the yield limit versus shear banding is considered, and, in particular, state of the art regarding the yield condition for modern materials is the subject of thorough study. Viscoplasticity models accounting for shear banding with related examples are under investigation in Chapter 8. The conclusions and remarks concerning further possible studies are provided in Chapter 9.

Acknowledgements

Many friends and coworkers supported and helped the author pursue this complex never-ending story on multiscale deformation mechanisms of different hard deformable metallic solids that I would like to recount, at least partly. As mentioned above, Andrzej Korbel, a Polish Academy of Arts and Sciences member, and Włodzimierz Bochniak became 'spiritus movens' of my long-time activity in this field. It happened due to the help of Mrs Romana Ewa Śliwa, a professor at the Rzeszów University of Technology. She was the first to see my preliminary presentations on localisation phenomena long ago, wisely suggesting contacts with already-knowledgeable and experienced material science researcher Andrzej Korbel. Then, during many years of my works on shear banding phenomena, it was Zdzisław Nowak, PhD, DSc, who showed me the possibilities of numerical analysis of plastic deformation processes accounting for the shear bands effects and identifying the shear banding contribution function. Also, Katarzyna Kowalczyk – Gajewska, PhD, DSc, gave me a helping hand in the numerical simulations of experimentally realised channel-die compression and shearing processes. The studies showed a valuable perspective with the significant experience and knowledge of professor Zenon Mróz, a Member of the Polish Academy of Sciences, and Katarzyna Kowalczyk – Gajewska on cyclically loaded tubes on the KOBO processes. I also received a big help from Mrs Aleksandra Manecka – Padaż, MSc, in elaborating the book's graphics. Together with discussions *in statu nascendi* of the work, her contribution became invaluable. The late professor Piotr Perzyna, my PhD advisor and scientific tutor, contributed to my studies with many valuable discussions about the shear banding model and its applications in the studies of viscoplastic processes.

References

Asaro, R.J. and Lubarda, V.A. (2006). *Mechanics of Solids and Materials*. Cambridge, New York: *Cambridge University Press*.

Bingham, E.C. (1916). An investigation of the laws of plastic flow. *US Bur. Stand. Bull.* 13: 309–353.

Bochniak, W., Korbel, A., Szyndler, R. et al. (2006). New forming method of bevel gears from structural steel. *J. Mater. Process. Technol.* 173: 75–83.

Bochniak, W., Korbel, A., Ostachowski, P., and Pieła, K. (2011). Superplastic flow of metal extruded by KoBo method. *Mater. Sci. Forum* 667–669: 1039–1044.

Bochniak, W., Korbel, A., Ostachowski, P. et al. (2013). Extrusion of metals and alloys by KOBO method, (Wyciskanie metali i stopów metodą KOBO). *Obróka Plastyczna Metali, (Metal Forming)* XXIV: 83–97, bilingual quarterly.

Frąś, L. (2015). The Perzyna viscoplastic model in dynamic behaviour of magnetorheological fluid under high strain rates. *Eng. Trans.* 63: 233–243.

Frąś, L. and Pęcherski, R.B. (2018). Modified split Hopkinson pressure bar for investigations of dynamic behaviour of magnetorheological materials. *JTAM* 56: 323–328.

Gurtin, M.E., Fried, E., and Anand, L. (2009). *The Mechanics and Thermodynamics of Continua*. Cambridge, New York: *Cambridge University Press*.

Korbel A. and Bochniak W, Method of plastic forming of materials, US Patent No. 5,737,959 (1998). European Patent No. 0711210 (2000).

Korbel, A. and Bochniak, W. (2003). KOBO type forming: forging of metals under complex conditions of the process. *J. Mater. Process. Technol.* 134: 120–134.

Korbel, A. and Szyndler, R. (2010). The new solution in the domain of metal forming – contribution of the Polish engineering idea, (Innowacyjne rozwiązania w obszarze obróbki plastycznej – udział polskiej myśli technicznej). *Obróbka Plastyczna Metali (Metal Forming)* XXI: 203–216, bilingual quarterly.

Korbel, A., Bochniak, W., Ostachowski, P., and Błaż, L. (2011). Visco-plastic flow of metal in dynamic conditions. *Metall. Mater. Trans. A* 42A: 2881–2897.

Perzyna, P. (1963). The constitutive equations for rate sensitive plastic materials. *Q. Appl. Math.* 20 (4): 321–332.

2

Physical Basis

2.1 Introductory Remarks

The plastic deformation of metals causes the evolution of microstructure. One assumes that the structural changes relate to the breading, mobility, the annihilation of dislocations, and grains' partitioning. One knows commonly from the literature on rational mechanics of materials, cf., e.g. Perzyna (1971), or for instance Nemat-Nasser (2004), the attempts of accounting for these processes on metallic solids' behaviour by supplementing thermomechanical description with the evolution equations for internal variables typically representing the effects of equilibrium between plastic hardening and softening produced by metal structure recovery. Such a view follows the premise that the mentioned structural changes generally provide metallic body global mechanical properties. However, it appears that it is not a satisfactory explanation of all vital aspects of material behaviour.

A good example is the hardening curve, showing a stepwise shape with several saturation and steady-state flow ranges accompanied by advanced plastic deformation, e.g. Richert and Korbel (1995). One reads in p. 339: 'The obtained results show that the strain localisation in the long-range shear bands, running through the whole volume of the samples, is the main deformation mechanism in AL 99,992 after a strain of about 4, exerted by the CEC method. The most important feature of these bands is their deep penetration through the structure (Fig. 5) and, consequently, the possibility of structural change in a wide area of a sample. The homogeneous structure and substructure are formed by mutual crossing of long-range shear bands and microbands (Fig. 6, 7, 10). Homogenisation is probably responsible for the onset of plastic flow at the same level of stress. This phenomenon may also be due to a change in the strain path as the samples were prestrained by the CEC method and next compressed'. CEC means cyclic extrusion

Viscoplastic Flow in Solids Produced by Shear Banding, First Edition. Ryszard B. Pęcherski.
© 2022 John Wiley & Sons Ltd. Published 2022 by John Wiley & Sons Ltd.

compression (*my comment*). The pictorial diagram in Fig. 12 of Richert and Korbel (1995) illustrates a strain-hardening curve development in the range of extensive deformations with the corresponding background of shear banding evolution. It shows an essential change of view in structural development issues accompanying advanced plastic flow. In the light of numerous experimental studies and the complex metallographic and microscopic observations with the application of different techniques for various metals and alloys, a new opinion is transpiring. It means that in the case of large plastic deformations, the plastic flow is realised by the hierarchy of shear banding in the microscopic, mesoscopic, and macroscopic levels. Also, for small or moderate strains preceded by the deformation path or variation of the loading scheme, plastic flow results in shear bands' multiscale hierarchy: micro-shear bands, clusters of micro-shear bands, and macroscopic shear localisation zones.

This chapter contains the synthetic approach to conclusions and observations, which helps formulate the viscoplastic flow description in metallic solids produced by shear banding. These views are underlined in the text uniquely, for they contain the physical motivation necessary to formulate the theoretical model. In particular, the set of observations denoted Observations 2.1, 2.2, . . . 6.1 includes the results of own inquiries. The heuristic foundations of the theoretical description of large inelastic deformations create the rational formulation of a multiscale system of shear banding.

2.2 Deformation Mechanisms in Single Crystals

This paragraph considers plastic glide and twinning as underlying plastic deformation mechanisms in crystals with the more detailed elucidation of inelastic slip processes. Then localised forms of plastic strain and the physical nature of shear banding are at the heart of the discussion.

2.2.1 Plastic Glide and Twinning

Schmid and Boas (1935) classical monograph deals with foundations of plastic deformation of single crystals and polycrystalline metallic solids. The textbooks, containing the necessary information for introducing material sciences and metallurgy of Dieter (1961, 1988) and Honeycombe (1984), as well as the overview works of Asaro (1983)), Neuhäuser (1983), and Basinski and Basinski (1979), consider and develop the similar problems of plastic slip and twinning. In the light of the discussion of numerous investigations, the cold plastic deformation of metals, i.e. in the temperature range below half of melting temperature, is realised by glide (slip) and twinning. The slip mechanism produces the displacement of one part of the crystal relative to the adjacent part, keeping both parts' crystallographic structure unchanged. The slip region delimits the thin layer of the node's thickness of the crystal lattice order. The explanation from a simple geometrical perspective gives (Dieter 1961), p. 82: 'X-ray diffraction analysis shows that the atoms in a metal crystal are arranged in a regular, repeated three-dimensional pattern. The atom arrangement of metals is most simply portrayed by crystal lattice in which the atoms are visualised as hard balls located at particular locations in a geometrical arrangement'. The layer remains parallel to the specific crystallographic plane called the slip plane. Since the slip occurs only on the defined slip planes that lie apart by specified distances, plastic strain's fundamental nature becomes discrete and heterogeneous.

On the other hand, the twinning mechanism also displaces two crystal parts along specific planes and directions. However, the two adjacent regions of a crystal become the mirror reflection of each other. Using the analogy of the shearing of a deck of playing cards of Gilman (1960), p. 189, we can consider the two modes of plastic deformation called translation-gliding and twin-gliding (twinning). A crystal structure shears into a mirror image of itself. The twin-gliding requires the same amount of shear at each atomic level of a suitably oriented crystal structure.

It is characteristic that plastic glide in a crystal does not appear simultaneously in all crystallographic planes and directions. It seems gradual instead if the slips activate sequentially in the crystallographic planes and directions that are the most suitable for applied loading. The maximum shear stress attains the critical value. It is expressed by the known Schmid criterion of plastic slip, cf., e.g. Hill (1966) and Phillips (2001):

$$\boldsymbol{b}\sigma\boldsymbol{n} = \tau_{cr}, \tag{2.1}$$

where $\boldsymbol{b}$ and $\boldsymbol{n}$ are the unit vectors defining the glide direction and the normal to the slip plane, while σ denotes the Cauchy stress tensor. The critical value of resolved shear stress τ_{cr} determines the plastic glide in a given slip system $(\boldsymbol{b}, \boldsymbol{n})$. The resolved shear stress in the considered slip system is called the Schmid stress. In particular, for the stretched single crystal of the cross-section area A with the application of tensile load P, the plastic slip criterion takes the form, cf., e.g. Dieter (1961), p. 99:

$$\frac{P}{A}cos\varphi cos\lambda = \tau_{cr}, \tag{2.2}$$

where φ and λ denote the angle of inclination of the unit normal $\boldsymbol{n}$ of the slip plane and the slip direction $\boldsymbol{b}$ concerning the tensile axis. The orientation factor $cos\varphi cos\lambda$ appears in the literature as the Schmid coefficient. The plastic glide process in a crystal depends on the glide's direction and the slip plane's orientation, i.e. it becomes an anisotropic one.

It means that in individual crystallographic planes and directions, smaller values of the critical amount of resolved shear stress τ_{cr} than in the other slip systems are realised, and we distinguish the so-called easy-slip systems. The directions and planes of easy-slip systems are usually most densely packed by network atoms. It results from crystalline lattices with metallic bonds, the distances between two adjacent planes, most densely filling up with atoms, are more significant than between other planes. In this way, the interaction between atoms decreases, and then the plastic slip requires smaller energy, i.e. more figuratively lower strength to overcome. The directions of an easy-slip system in a crystal are the shortest distances at which translation restores the original crystal lattice due to a unit shift. One should add that slips may also occur when slips in easy systems become blocked in the case of advanced deformations. An example of such a situation is an array of micro-shear band formation processes, which means shear banding. This question will be the subject of further analysis. The phrase 'easy slip systems' introduces Korbel to underline that under certain conditions, such as the propagation of micro-shear bands, plastic slip may also occur in other crystallographic systems, so-called difficult slip systems. Such particular slip systems are colloquially referred to as 'non-crystallographic', cf., e.g. Korbel (1990a,b) and Korbel and Bochniak (1995).

2.2.2 Hierarchy of Plastic Slip Processes

The results of plastic slips reveal the change of shape, distortion of the crystalline body. Considering the metal sample's properly polished surface, one can observe the characteristic traces of steps after deformation. Investigations of such steps using optical microscopy show them as a contrast in lines and bands. One of the first original observations of deformed metal's surface belongs to Ewing and Rosenhain (1900). These researchers coined the widely used concepts and terms of 'plastic steps', 'slip lines', and 'slip bands' cf., e.g. Asaro and Lubarda (2006, pp. 502–503). A more detailed discussion and citation of the original paper of Ewing and Rosenhain (1900), p. 503: 'When the metal is strained beyond its elastic limit, as say by a pull in the direction of the arrows, yielding takes places by finite amounts of slips at a limited number of places, . . . This exposes short portions of inclined cleavage or gliding surfaces, and when viewed in the microscope under normally

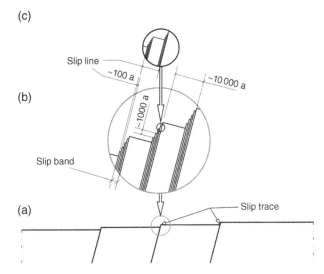

Figure 2.1 An overview diagram of the multiscale hierarchy of plastic slip processes: (a) characteristic traces of slip visible as slip steps on the hypothetical cross-section of the polished surface of the plastically deformed metal, (b) magnified picture of a single slip step showing the schematic structure of slip band and undeformed regions between the bands, (c) a zoomed picture of slip line revealing the discrete structure of plastic glide. *Source:* Copyright of Aleksandra Manecka – Padaż.

incident light these surfaces appear black because they return no light to the microscope'. Ewing and Rosenhain (1900) results and the discussion in Dieter (1961) lead to the idea of the schematic diagram of the multiscale hierarchy of plastic slip processes displayed in Figure 2.1. The parameter 'a' scaling the different magnifications of the structure of slip lines, slip bands, and slip steps, respectively, corresponds to the lattice parameter of a given metal crystal, e.g. $a = 3.597$ Å for Cu or $a = 4.046$ Å for Al, where 1 Å $= 0.1$ nm, while 1 nm equals 1.0×10^{-9} m.

The optical microscopy or even the naked eye's observations show that slip bands reveal plastic deformation in thin layers with undeformed metal zones in between. The transmission and scanning electron microscopy application shows that the thin layers are discrete due to slips generated by massive breeding, the spread, and the propagation of dislocations. Due to the solid long-range interactions, the entire ensemble of dislocations becomes organised in space and time as high-energetic structures. The high-energy groups of dislocations transform in the dislocation systems of lower internal energy. They are visible using transmission electron microscopy in dislocation loops and dislocation tangles and single or double dislocation walls being the elements of microbands, cells, or subgrains, e.g. Honeycombe (1984) as well as Kuhlmann-Wilsdorf (1976) and Kuhlmann-Wilsdorf (1985). The abovementioned results lead to a profound question about the conceptual gap between the multilevel geometry of slip traces depicted in Figure 2.1 and the physical reality of plastic deformation. It seems like a specific example of an interplay of platonic and Aristotelian concepts of Hellenistic science, recalling the masters from a distant era, cf. e.g. Russo (2004) and Rovelli (2011). Both authors brought us closer to Anaximander (c. 610–c. 546) BCE, the great thinker from Miletus, creating the 'inquiry into nature' and forming the foundations for today's scientific reasoning tradition and critical thinking. According to Rovelli (2011, p. 179): 'Scientific thinking is, therefore, a continuous quest for novel ways of conceptualising the world. Knowledge is born from a respectful but radical act of rebellion against what we currently think. This is the richest heritage the West has bequeathed to today's global culture, its finest contribution'.

Observe that, in our case, bridging the mentioned conceptual gap means feeling an arduous journey of dislocations towards the surface to leave the permanent trace of existence. Such a view is concordant with Phillips's witty statement (Phillips 2001), p. 366: 'An example of the emergence of slip steps as a result of plastic deformation as revealed using atomic force microscope is shown in Fig. 8.3. As will become more evident below, these lines are the point of exit of dislocations after their tortuous journey through the crystal's interior. Each dislocation carries with it an elementary unit of lattice translation (the so-called Burgers vector) which implies a relative displacement across a given plane by a length of order of the lattice parameter'.

More advanced plastic deformations relate to an increase of slip lines in individual slip bands, and also, the number of slip bands is growing. In such a case, the bands' undeformed regions are filling up with new secondary slip lines. These are the accommodation slips produced by micro-stresses that appear as an equilibrating external bond reaction (grips). Increasing values of micro-stresses lead to Schmid's increase in the critical amount of the resolved shear stress in different orientation systems than the dominant primary slip system's orientation. In this way, the plastic slips in the primary system activate, and the rotations of crystalline lattice produced by the grip bonds become irreversible. According to the observations discussed by Neuhäuser (1983), the resulting slip lines organise clusters of the thickness from hundreds to several thousands of the lattice parameter a. Such clusters are called *coarse slip bands*. The above discussion leads to the following profound observation:

Observation 2.1

The coarse slip bands stand out among other structural heterogeneities with the following features:

- they are parallel to slip lines and can develop within the whole interior of a single crystal or a single grain in a polycrystal
- propagation range is much longer than slip bands and often appears along with the entire grain or a single crystal
- the plastic slips are concentrated in coarse packets that stand out from the surrounding structure.

The coarse slip bands are separated by zones of one to several hundred microns thick in which accommodation slips may occur. Such a hierarchy of slip processes in easy-glide systems transpires on the surface of a single crystal or a separate grain of polycrystalline metal subjected to monotonic loading. Micro-bands as clusters of slip lines and coarse slip bands are a qualitatively new element of the said hierarchy of slip processes representing one form of localisation of plastic deformation, cf. Figure 2.1.

2.2.3 Localised Forms of Plastic Deformation

Almost even distribution of the slip line over the sample's gauge length, typical for the second linear stage of strengthening, is heterogenised when passing to the third, 'parabolic' stage of the stress–strain curve. One can observe the localisation of plastic deformation in the shape of coarse slip bands.

The further step of the crystal strain may lead to the localisation in necking or macroscopic shear bands. For example, pure metal crystals are characterised by an extensive necking in the tensile test, which develops gradually and only at the final stage passes into a localised shear zone preceding the plastic separation by slide. Early results of experimental observations indicate this effect, cf., e.g. Taylor and Elam (1925) and Puttick (1963) carried for Al single crystals, and also Göler and Sachs (1930) and Saimoto et al. (1965) for Cu single crystals, as well as Sachs and Weerts (1930) for crystals of alloy Au–Ag.

In contrast, the strengthened alloy crystals, e.g. Guinier–Preston zones or coherent precipitates, are characterised in that macroscopic shear bands initiate without visible necking. In this case, the plastic deformation's localisation in the form of a macroscopic shear band leads directly to plastic separation by slide. It is confirmed, for example, by early studies of Elam (1927), Karnop and Sachs (1928), and Beevers and Honeycombe (1959), as well as Price and Kelly (1964) performed for Al alloy crystals. Effects of strengthening in Cu–Zn crystal alloys (α brass) and pioneering studies on the impact of slip sequences on plastic strengthening, localisation of plastic deformation, and plastic failure were carried by Masima and Sachs (1928) as well as by Göler and Sachs (1929). Spitzig (1981) investigations and Dève et al. (1988) for single-crystal alloys Fe–Ti–Mn and Fe–Ti indicate a strong dependency on the form of plastic localisation on the initial orientation and consequently on the geometry of active-slip systems. When active-slip systems leading to a plane state of deformation were developed in the monotonic tensile test, the observed form of localisation was the macroscopic shear band. However, when the crystal's initial orientation did not lead to a plane deformation, the localisation's predominant form was necking followed by plastic separation.

The localisation of plastic deformation in macroscopic shear bands met with great interest in materials science and mechanical engineering specialists dealing in the plastic deformation of metals. It became the subject of interdisciplinary studies. It is worth noticing that shear bands formed during the deformation process are not only a precursor of ductile separation of metallic solid, but also they are becoming the dominant mechanism of advanced plastic deformation. It is necessary to know about the initiation and development of shear bands in single crystals to explain the complex mechanisms of shear band formation in polycrystals. Experimental research on single crystals subjected to various deformation processes, using many techniques for observing the sample surface and its internal structure utilising optical and electron microscopy and X-ray methods, provide many theoretically essential results. They help answer the questions about the physical nature of shear bands and their formation mechanisms. Shear bands formed in monotonically stretched monocrystals of Al–Cu alloys were studied by Chang and Asaro (1981), Szczerba and Korbel (1987), and Korbel and Szczerba (1988). A rich set of observation results of shear bands in stretched Mo–C single crystals was discussed in the monographic work (Luft 1991), which also added a very extensive bibliography of the subject. Pivotal research results regarding the initiation and evolution of shear bands in stretched Cu single crystals were thoroughly analysed and discussed, e.g. by Dubois (1988) and Dubois et al. (1988) as well as Yang and Rey (1993, 1994). In the tensile test, the mechanism of forming a single shear band or a system of two intersecting shear bands is visible, which in reality always has a sequential character. In the rolling test and the channel-die test (constrained compression),

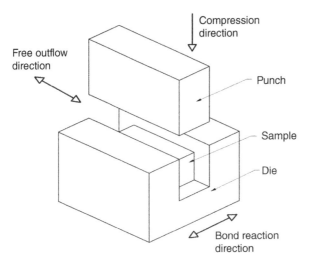

Compression
direction

Free outflow
direction

Punch

Sample

Die

Bond reaction
direction

Figure 2.2 Schematic view of channel-die test idea. *Source:* Copyright by Aleksandra Manecka – Padaż.

Figure 2.2, on the other hand, the material behaviour at advanced plastic deformations can be the subject of examination when shear bands become the dominant mechanism. Havner's monograph (Havner 1992) is worth mentioning as it contains the comprehensive historical survey of early original axial-load experiments and latent hardening in single crystals. A separate chapter in Havner's (1992) work is devoted to the detailed exposition of the first works on crystals analysis in a channel-die compression. The results of the tests of crystals subjected to the rolling process presented, for example, for Cu: Korbel et al. (1986), Bochniak (1988, 1989), and Wróbel et al. (1990), as well as, for Cu–6%Al: Embury et al. (1984) and Korbel et al. (1986) and single crystals of austenitic steel (Wróbel et al. 1995). However, the observation results of shear bands in single crystals subjected to constrained compression in the channel-die test are presented by Harren et al. (1988) for Al–3%Cu and Jasieński and Piątkowski (1988, 1993) for Cu.

2.2.4 Physical Nature of Shear Bands

Analysing the wealth of information from previous experimental works and interpreting the results of metallographic observations lead to a coherent picture of shear bands' formation and evolution in single crystals. The resulting view takes the form:

Observation 2.2

Macroscopic shear bands in a single crystal are complex and multiscale structures created due to the organisation's time and space of almost parallel layers of shearing zones. Visible shear bands have a range corresponding to the area covering the entire width of the sample. Such a fundamental element is an active micro-shear band with a constant thickness of order 0.1 μm.

Micro-shear bands in a single crystal are crystallographic, i.e. the plastic glide continues further in easy-slip systems. A rectilinear course characterises them through the existing substructure and sub-grain boundaries showing characteristic faults at the intersection points. The faults indicate significant shear strains, generally ranging from 1 to 10, carrying by micro-shear bands. The shear band's evolution consists of successive generations of active micro-shear bands about 0.5 μm apart, propagate along with the distance of the order of 50 μm, and contribute to the total shear strain and the increase in the active part of the shear band. They saturate, remaining behind the expanding shear band as a passive element of the deformed crystal structure. For example, it is confirmed by the results of in situ observations of crystal deformations using a scanning electron microscopy (SEM) obtained by Yang and Rey (1983, 1984).

Visible on the single-crystal sample's lateral surface, traces of planes limiting the macroscopic shear band may deviate from traces of slip. This deviation can be from a few to several degrees, depending on the shear strain difference in the band relative to the adjacent material. It may change during the development of the shear band. Research by Dève et al. (1988) shows that in the case of an Fe–Ti–Mn extended crystal, the shear strain in the band is of the order of 3, and in the surrounding material is about 0.1. Despite such large deformation gradients, the material maintains its continuity.

X-ray observations also show the crystal lattice's misorientation within the shear band relative to the adjacent material's lattice orientation. The lattice rotation in the band increases with an increase in the shear strain, causing the slip planes to be arranged almost parallel to the plane, limiting the shear band's active part. It confirms the hypothesis of the crystallographic nature of shear bands. It means that the plastic deformation in the shear band still occurs in an easy-slip system, which is much more intense due to geometric softening caused by proper rotation of the slip plane, creating a local increase in the Schmid coefficient. This interpretation is possible only when we have to do with a sample with a rectangular cross-section. A single or intersecting shear band planes are perpendicular to the observed surface in the gauge volume sample. It corresponds to the often-appearing situation when the crystal deforms in a plane state of deformation. The above discussion finds confirmation in the results of observations in the works of Harren et al. (1988; Dève et al. (1988). One should add that new slip systems may activate though they are not visible on the sample's observed surface during such advanced deformations. These systems' operation often manifests itself because the visible traces of shear band boundaries or micro-shear bands locally undergo additional shifts and rotations, obscuring and sometimes even falsifying information about their orientation. One should not overestimate the importance of geometric softening due to the shear band's crystal lattice rotation. Observations on the micro-shear bands show substantial heterogeneity in the crystal lattice orientation and the band, cf., e.g. Korbel (1985).

The formation of dislocation junctions, jogs, dislocation debris, and loops, as well as clusters of tangled dislocations, appears in the course of the plastic deformation process. *Seeing is believing* – I still remember, May 1980, the *in situ* observations with the use of high-voltage transmission electron microscopy of dislocations breading and mobility processes due to the kind invitation of professor Toru Imura from Nagoya University and professor Tatsuo Tabata from Osaka University, cf. the more detailed description in Pęcherski (1983).

These structural elements contribute to the material strain hardening. The development of three-dimensional networks of dislocation tangles and cells accommodates the large strains. Incorporating the dislocations by traversing the cell's interior into the cell wall produces rigid relative rotation concerning its neighbourhood. At an earlier deformation stage, the average cellular misorientation is relatively tiny, and the glide dislocations can cut through nearly all cell walls. At higher strains, however, some cell walls become impenetrable and form a new substructure. The network of impenetrable walls separates volumes of high lattice misorientation. This process of a new subdivision corresponds to fragmentation of the crystalline lattice, and the volume elements are fragments. The average diameter of these fragments diminishes in the strain and the misorientation angle, cf. Rubtsov and Rybin (1978), Langford and Cohen (1975), Gardner et al. (1977), and Gardner and Wilsdorf (1980).

The development of the fragmented structure with high lattice misorientation makes the onset of shear banding and initiation of ductile fracture (Chang and Asaro 1980; Wilsdorf 1980; Orlov and Shitikova 1981). The effect of 'local lattice rotations' within shear bands is considered

(Asaro 1979). One of the fragmentation's primary ways develops through band structures by propagating from the cell or fragment boundary into the interior. Of great importance here, as noted by Vergazov et al. (1977), is that the neighbouring edges of the bands terminate abruptly within the cell or fragment interior and turn the crystal lattice by equal angles of opposite sign. Similar phenomena are analysed (Kocks et al. 1980). They discuss pairs of sub-boundaries of opposite signs producing large lattice misorientation. The material volume containing such pairs of polarised sub-boundaries plays the role of small 'nuclei' from which 'shear bands' observed in free compression and rolling might originate. Also, Chandra et al. (1982) report that unstable oriented crystals reveal fragmented bands of high misorientation.

The mechanisms, as mentioned earlier for the formation of terminated bands and polarised pairs of sub-boundaries and fragmented bands, are similar and find interpretation through some cooperative forms of highly concentrated dislocation motion and the onset of collective degrees of freedom. It produces the local rotations of material volumes that eventually lead to theoretical visualisation as partial disclination dipoles and loops, cf. Rubtsov and Rybin (1978), Rybin (1978), and Orlov and Shitikova (1981) as well as Romanov and Vladimirov (1983). These authors consider the formation of partial disclination dipoles and loops as a critical state model, very characteristic of the onset of plastic flow localisation in the form of shear banding. On the other hand, the importance of inhomogeneous lattice rotations for the start of necking and shear band formation in single crystals is underlined (Saimoto et al. 1965; Chang and Asaro 1981; Lisiecki et al. 1982); see also Pęcherski (1983, 1985) for more detailed review of microscopic observations.

The question arises: what is the mechanism of the formation of micro-shear bands in the crystal?

The first works on observing shear bands in stretched Al–Cu crystals (Price and Kelly 1964; Chang and Asaro 1981) noted that micro-shear bands' precursors are coarse slip bands. However, then there was no clear view of the conversion of coarse slip bands into micro-shear bands. The coarse slip bands were treated instead as some imperfection inducing the shear band. Simultaneously, the crystal shear band found interpretation regardless of the slip mechanism in an easy system, emphasising the deviation of traces of the shear band border visible on the sample surface from the traces of slip lines, considering it as evidence of non-crystallographic nature of the shear band. More recent studies discussed above confirmed the hypothesis of the crystallographic nature of micro-shear bands. Therefore, it seems that micro-shear bands have developed from those coarse slip bands of more favourable conditions for developing much larger shear strains than in the others. The following elastic unloading outside the shear band can enhance plastic flow in the band. The reaction of grip constraints causes the entire band to rotate relative to the stretching axis and the crystalline lattice's disorientation. In summary, one can say that the mechanism of micro-shear band initiation in single crystals is directly related to the formation and development of coarse slip bands. The same holds for polycrystals.

2.3 Plastic Deformation in Polycrystals

2.3.1 Mechanisms of Plastic Deformation and the Evolution of Internal Micro-Stresses

Observing with an optical microscope, after polishing and etching the deformed polycrystalline sample's surface, one can visualise a characteristic topography in the form of traces of straight lines and slip bands that have a similar orientation within a single grain. In the initial plastic deformation process, groups of almost parallel slip bands are visible in only individual grains most favourably oriented relative to the applied load. As load increases, slip bands appear in a continually increasing number of grains. With even more significant deformation, intersecting bands perform, indicating the launch of new slip systems, contributing to the increasing violation of these bands' straightness. It results from increasing structural defects and the transition from gliding in one plane system to another arrangement. The slip bands divide the grain into individual parts, which, as deformation develops, rotates together with the crystal lattice relative to the set reference system and changes its shape. It launches new slip systems, divides the grain into sub-grains, and reshapes the grain and its dimensions. Most often, grain elongation is visible along the direction of plastic flow. The amount of change in the preferred orientations, shape, and sizes of the different grains of the deformed polycrystal is different due to the various lattice directions relative to the applied load. These effects refer to texture evolution, which may characterise preferred grain orientation or the nature of morphological texture resulting from grains' shape and arrangement. Similar to the plastic deformation of single crystals, the metallic material structure's local rotation significantly impacts plastic deformation-induced anisotropy formation and evolution. Summing up the above discussion, the following observation arises:

Observation 2.3

The local orientations of the metallic material structure appear in reality as the rotations of polycrystalline micro-volumes with the embedded crystal lattice due to complex plastic deformation processes resulting from the applied load and external constraints produced by boundary conditions adjacent grains' interaction due to the surrounding material's reaction.

There is an inhomogeneous distribution of plastic deformation inside and around grains of polycrystalline material. This heterogeneity is the cause of self-equilibrating micro-stresses that remain after the unloading of the body.

The micro-stresses change during the plastic deformation process. Their impact on the yield strength and strengthening of the material also changes. Micro-stress's macroscopic effect during cold plastic deformation changes strained metal's stored energy, mainly in its dislocation structures. It is the part of the plastic deformation process's mechanical energy that does not dissipate in heat. Taylor and Quinney (1934) have already dealt with the issue of stored energy. The monograph of Bever et al. (1973) is devoted to an extensive study of this subject, which contains thermodynamic foundations, a comprehensive discussion of measurement methods together with an in-depth analysis of the impact of various parameters, supported by a review of dislocation models and a rich bibliography of the subject. Raniecki (1984) emphasised the importance of including stored energy to describe plastic strengthening within thermoplasticity – the field of materials' rational mechanics, cf. Perzyna (2012) for the theoretical framework.

The original method of measuring and analysing the change in energy stored while stretching the sample using the thermovision technique was presented (Oliferuk et al. 1995). Recent experimental results regarding energy storage capacity in deformed metal combined with metallographic observations under the optical microscope of slip lines and micro-shear bands on the surface of specially polished samples were presented by Oliferuk (1995) and Oliferuk et al. (1996, 1997). It transpires that when an increase in micro-stresses accompanies the increasing deformation in the early stage of plastic flow, it also increases energy storage capacity. Observations of the samples' surface show that this process can occur until secondary slip systems are activated, and micro-shear bands' onset appears.

The graph of energy storage capacity as a function of deformation then reaches its maximum. A further increase in plastic deformation is accompanied by an apparent decrease in energy storage capacity and increased heat emission generated by dissipative processes of slip and propagation of micro-shear bands. Then, under conditions of dynamic accumulation of dislocations at grain boundaries, a partial relaxation of micro-stresses occurs, and favourable conditions arise for propagating micro-shear bands. As deformation increases, the energy storage capacity decreases, and micro-shear bands' massive participation is increasingly visible. The cited results of experimental observations are summarised in the following view, which will be a vital element of the theoretical model's physical motivation.

Observation 2.4

Micro-shear bands' massive formation and propagation mechanisms for a specific loading path correlate with the saturation stage of micro-stresses' growth and their partial relaxation.

Changing the loading path or load pattern will cause the mentioned saturation effect for another micro-stress state. It happens since the mechanical state in which micro-shear bands develop depends on the load path. It is the main difficulty in describing the impact of micro-shear bands on the plastic deformation of metals. This challenge is a matter of discussion below.

2.3.2 Micro-shear Bands Hierarchy and Their Macroscopic Effects

Large plastic strains of polycrystalline metals and alloys often appear in highly constrained conditions, such as in technological processes of plastic shaping or necking operation. It can lead to a ductile failure or transform into an intense concentration in intensive plastic shear layers called shear bands. As in single crystals, experimental studies of polycrystals show that micro-shear bands occur at different observation levels. The shear layers appear in correspondingly small areas but cover more than a few grains we call micro-shear bands, while the layers that occur across the entire sample are referred to as macroscopic shear bands.

The first microscopic observations of plastic deformations, accompanied by shear bands, were published by Adcock (1922). Adcock studied the samples of Cu–20Ni alloy rolling them to reduce 50% of the cross-section. Then he explored the sample's side surface using the optical metallography technique and registered dark etched band-shaped traces that, fixing their initial direction, run through many grains and are inclined to the rolling surface at an angle (25°–40°). Such marks were also visible in other metals and alloys. Over time, in literature, they got the original name *shear bands*.

The subsequent observations of shear bands for various metals and alloys became the discussion subject (Sevillano et al. 1982). An exhaustive bibliography and a large number of experimental results obtained using different techniques for different metals are also provided in the works of Duggan et al. (1978), Hatherly (1983), and Malin and Hatherly (1979), as well as Hatherly and Malin (1984), Korbel et al. (1986), and Korbel (1987), and Deve and Asaro (1980). Also, the papers of Bochniak (1989) and Dybiec et al. (1989) are worth mentioning. From the vast amount of information essential for specialists in metallographic research, materials science, and structural transformations accompanying large metal deformation, we will choose those relevant for the macroscopic description of the impact of shear bands on the change mechanical properties of the material. We will use a synthetic discussion of the most important experimental results to apply the phenomenological description of plastic deformation of metals with shear band effects included in the author's works: Pęcherski (1991, 1992, 1993).

According to the terminology adopted in the literature, cf., e.g. Hatherly (1983) and Korbel et al. (1986) as well as Korbel and Martin (1986, 1988):

Micro-shear bands in polycrystals are visible as long and thin layers of material with a constant thickness of 0.1 µm, characterised by intense plastic shearing. These layers cross grain boundaries without deviations and form a spatially organised family with specific geometry relative to the strain tensor's principal axes. Micro-shear bands carry extensive shearing and lie in planes that usually do not coincide with the crystalline planes of the easy slip systems in the grains they pass through.

The micro-shear bands defined in this way are an elementary, active component of macroscopic shear bands. An analysis of their structural and morphological characteristics leads to the following observation.

Observation 2.5

Macroscopic shear bands in polycrystals are hierarchical structures resulting from the time and space organisation across the scales of the almost-parallel shearing layers. Shear bands have a range corresponding to the macroscopic areas of the deformed body plastically.

Experimental studies show that micro-shear bands act only once and propagate at very high speeds in many grains areas. Then, they no longer contribute actively to plastic deformation, becoming a passive element of the structure. Observations also show that active micro-shear bands' contribution to the total plastic deformation generally increases with plastic flow progression. However, it changes very irregularly during the process, cf., e.g. Malin and Hatherly (1979), as well as Hatherly and Malin (1979, 1984) and Yeung and Duggan (1987). Also, information about the deformation at which the first micro-shear bands appear is very fragmentary and incomplete.

Further systematic and comprehensive experimental studies are needed to combine microscopic observations of micro-shear bands on the samples' surface and measure and analyse the change of energy stored in deformed metal, such as the thermovision technique used cf., e.g. Oliferuk et al. (1996). Also, Pieczyska (1999) shows that using the thermovision observations to measure the piezocaloric effect appears helpful in developing a method for determining the yield point. The situation is complicated because micro-shear bands' initiation depends on the deformation path or load pattern. It is confirmed, for example, by the results of the channel-die test (Figure 2.2), for polycrystalline copper obtained by Bronkhorst et al. (1992), Anand and Kalidindi (1994), and Paul et al. (1996). Studies conducted by Korbel (1987, 1990a,b, 1992) and Bochniak (1989) as well as by Korbel and Bochniak (1995) show, however, that micro-shear bands can be generated with any deformation by properly controlled deformation path change. It laid the foundation for new metal-shaping technologies developed by these authors. See the concept of the KOBO method discussed in the Introduction, Section 1.2.

Analysis of the experimental results shows that the active micro-shear bands, for example, are deflected by about $\pm35°$ relative to the rolling plane when rolling. Their surfaces are perpendicular to the side surface of the sample. However, one observes that the deviation angle values may vary in the range from 15° to 50°, cf., e.g. Hatherly and Malin (1984) and Korbel (1990a). It is worth emphasising that such a large dispersion of the deviation angle values may result from mismeasuring this angle. It is not easy to distinguish the newly created micro-shear bands from the traces of the earlier ones that rotate together with the material in the rolling direction. Similarly, in the process of stretching or compression, the planes of micro-shear bands form an average angle $\pm(38° \pm 2°)$ relative to the axis of maximum principal stress, cf. Anand and Spitzig (1980).

Observation 2.6

The sequences of successive generations of active micro-shear bands competing with plastic slip and twinning mechanisms control the visco-plastic flow.

Observation 2.7

A characteristic feature of the micro-shear bands' geometry pattern is the plane's slope representing all micro-shear bands' average orientation belonging to one shear system. This plane is associated with maximum shear stress and is deviated from it by a certain angle β, whose average value is in the range $(5° - 10°)$.

The above deviation is characteristic of micro-shear bands generated during deformation processes that occur under nearly isothermal conditions.

2.3.3   Physical Nature of Micro-shear Bands in Polycrystals

The physical mechanism of micro-shear band initiation in a polycrystal is not yet fully understood.

Nevertheless, the results of the experimental observations discussed earlier indicate that this mechanism is similar to the formation of a micro-shear band in a single crystal.

According to Korbel's hypothesis, which is confirmed by the results of numerous experimental studies, micro-shear bands develop from coarse slip bands, limited to a single grain. Therefore, they have a crystallographic origin associated with slips' activation and time and space in easy systems. Their further propagation is related to the relaxation of micro-stresses generated by the pile-up at the grain boundary in the dynamic dislocation groups that make up the coarse slip band. Ease of micro-stresses can cause easy-slip systems to run in the adjacent grain or, if the dynamic concentration of micro-stresses is high enough, activation of slips in one of the 'difficult slip systems'. Running the 'difficult slip systems' in several adjacent grains can promote coplanar slip propagation in the form of micro-shear bands, cf. Korbel (1985, 1990a,b). This hypothesis may also explain, in my opinion, the different behaviour of the so-called adiabatic shear bands.

2.3.4   Comments on 'adiabatic' Micro-shear Bands

The discussion of various forms of localisation of plastic deformation in poly-crystals would be incomplete if it would not add that the localisation in shear bands is also visible as thermal effects. At the same time, thermo-mechanical couplings play a crucial role. It is characteristic that they are then called 'adiabatic' shear bands. One should emphasise that these are 'adiabatic' effects that occur in macroscopic volumes of material. There is on this subject reach literature, cf., e.g. the monograph of Bai and Dodd (1992), also works of Duszek and Perzyna (1991), and Nguyen and Nowacki (1997), and Wright and Ravichandran (1997). Comprehensive information on structural observation results of 'adiabatic' shear bands contain monographs of Dodd and Bai (1987) and Meyers (1994). These bands are characterised because they lie in a plane approximately coinciding with the plane of maximum shear stress. Then the deviation angle is near $\beta = 0°$. This qualitative difference can be explained, according to the author, by the influence of micro-stresses, which play, according to Korbel's hypothesis, a critical role in the initiation and propagation of micro-shear bands. The micro-stresses disturb the state of stress induced locally by external loading. As a result, a critical coarse slip band will develop in an easy slip system that lies in a plane that deviated from the plane of applied shear stresses. It will then propagate through the adjacent grains in 'difficult slip systems', resulting in a 'non-crystallographic' micro-shear band that is spread along the plane that deviated a few or a dozen degrees from the plane of maximum shear stresses. In the case of 'adiabatic' shear bands, thermal effects can cause the impact of micro-stresses to diminish or even to disappear. At least qualitatively, it is confirmed by Gadaj et al. (1996) research.

References

Adcock, F. (1922). The internal mechanism of cold-work and recrystallization in Cupro—Nickel. *J. Inst. Met.* 27: 73–92.

Anand, L. and Kalidindi, S.R. (1994). The process of shear band formation in plane strain compression of fcc metals: effects of crystallographic texture. *Mech. Mater.* 17: 223–243.

Anand, L. and Spitzig, W.A. (1980). Initiation of localized shear bands in plane strain. *J. Mech. Phys. Solids* 28: 113–128.

Asaro, R.J. (1979). Geometrical effects in the inhomogeneous deformation of ductile single crystals. *Acta Metall.* 27: 445–453.

Asaro, H.J. (1983). Micromechanics of crystals and polycrystals. *Adv. Appl. Mech.* 23: 1–115.

Asaro, R.J. and Lubarda, V.A. (2006). *Mechanics of Solids and Materials.* Cambridge, New York: Cambridge University Press.

Bai, Y. and Dodd, B. (1992). *Adiabatic Shear Localization.* Oxford: Pergamon Press.

Basinski, S.J. and Basinski, Z.S. (1979). Plastic deformation and work hardening. In: *Dislocations in Solids*, vol. 4 (ed. F.R.N. Nabarro), 261–362. Elsevier, North-Holland, Amsterdam.

Beevers, C.J. and Honeycombe, R.W.K. (1959). Ductile fracture of single crystals. In: *Fracture* (ed. B.L. Averbach et al.), 474–492. New York: Wiley.

Bever, M.B., Holt, D.L., and Titchener, A.L. (1973). The stored energy of coldwork. *Progr. Mater. Sci.* 17: 5–192.

Bochniak, W. (1988). Strong localization of deformation in copper single crystals deformed at different temperatures. *Arch. Metall.* 33: 419–454.

Bochniak, W. (1989). Lokalizacja odkształcenia. Mechaniczne i strukturalne aspekty niestatecznego plastycznego płynięcia mono- i polikrystalicznej miedzi przy różnych temperaturach. Nadplastyczność w warunkach wysokotemperaturowego wymuszenia zmiany drogi odkształcenia. Zeszyty Naukowe AGH. Nr 122, Kraków (*habilitation thesis in Polish: Strain localization. Mechanical and structural aspects of instable plastic flow of mono- and polycrystalline Copper by different values of temperature. Superplasticity under the extorsion conditions of high temperature changes of deformation paths.*)

Bronkhorst, C.A., Kalidindi, S.R., and Anand, L. (1992). Polycrystalline plasticity and the evolution of texture in FCC metals. *Philos. Trans. R. Soc. London* A341: 443–477.

Chandra, H., Embury, J.D., and Kocks, U.V. (1982). On the formation of high angle grain boundaries during the deformation of aluminum single crystals. *Scr. Metall.* 16: 493–497.

Chang, Y.W. and Asaro, R.J. (1981). An experimental study of shear localization in aluminum-copper single crystals. *Acta Metall.* 29: 241–257.

Dève, H.E. and Asaro, R.J. (1980). The development of plastic failure modes in crystalline materials: shear bands in fcc polycrystals. *Metall. Trans.* 20A: 579–593.

Dève, H.E., Harren, S., McCullough, C., and Asaro, R.J. (1988). Micro and macroscopic aspects of shear band formation in internally nitrided single crystals of Fe-Ti-Mn alloys. *Acta Metall.* 36: 341–365.

Dieter, G.E. (1961). *Mechanical Metallurgy.* London: McGraw-Hill.

Dieter, G.E. (1988). *Mechanical Metallurgy. SI Metric Edition — Adapted by D. Bacon.* London: McGraw-Hill.

Dodd, B. and Bai, Y. (1987). *Ductile Fracture and Ductility with Applications to Metalworking.* London: Academic Press.

Dubois, Ph. (1988). Etude mistallographique de l'initiation et de la propagation de bandes de cisaillement dans les metaux purs. These prèsentè a l'Universitè Paris-Nord pour obtenir le grade de Docteur, Juin 1988.

Dubois, Ph., Gasperini, M., Rey, C., and Zaoui, A. (1988). Crystallographic analysis of shear bands initiation and propagation in pure metals. Part II Initiation and propagation of shear bands in pure ductile rolled polycrystals. *Arch. Meth.* 40: 35–40.

Duggan, B.J., Hatherly, M., Hutchinson, W.B., and Wakefield, P.T. (1978). Deformation structures and textures in cold-rolled 70:30 brass. *Met. Sci.* 12: 343–351.

Duszek, M.K. and Perzyna, P. (1991). The localization of plastic deformation in thermoplastic solids. *Int. J. Solids Struct.* 27: 1419–1443.

Dybiec, H., Rdzawski, Z., and Richert, M. (1989). Flow stress and structure of age-hardened Cu-0.4wt%Cr alloy after large deformation. *Mater. Sci. Eng.* A108: 97–104.

Elam, C.F. (1927). Tensile tests on alloy crystals. *Proc. R. Soc.* 115A: 133–169.

Embury, J.D., Korbel, A., Raghunathan, V.S., and Rys, J. (1984). Shear band formation in cold rolled Cu-6% Al single crystals. *Acta Metall.* 32: 1883–1894.

Ewing, J.A. and Rosenhain, W. (1900). The crystalline structure of metals. *Philos. Trans. R. Soc. London* 193: 353–375.

Frhr. Von Göler and Sachs (1929), Zugversuche an Kristallen ausKupfer und α-Messing. Z. Phys.55: 581–620.

Gadaj, S.P., Nowacki, W.K., and Pieczyska, E.A. (1996). Changes of temperature during the simple shear test of stainless steel. *Arch. Mech.* 48: 779–788.

Gardner, R.N. and Wilsdorf, H.G.F. (1980). Ductile fracture initiation in pure α-Fe. Part I. Macroscopic observations of deformation history and failure of crystals. Part II. Microscopic observations of an initiation mechanism. *Metall. Trans.* 11A: 653–669.

Gardner, R.N., Pollock, T.C., and Wilsdorf, H.G.F. (1977). Crack initiation at dislocation cell boundaries in the ductile fracture of metals. *Mater. Sci. Eng.* 29: 169–174.

Gilman, J.J. (1960). Physical nature of plastic flow and fracture, Plasticity, Proc. of the Second Symp. on Naval Structural Mechanics (ed. E. H. Lee and P.S. Symonds), Pergamon Press, 43–99, Oxford, London, New York, Paris.

Göler, Frhr.Von and Sachs, G. (1929). Zugversuche an Kristallen aus Kupfer und α-Messing. *Z. Angew. Phys.* 55: 581–620.

Harren, S.V., Dève, H.E., and Asaro, R.J. (1988). Shear band formation in plane strain compression. *Acta Metall.* 36: 2435–2480.

Hatherly, M. (1983). Deformation at high strains, strength of metals and alloys. In: *Proceedings of the 6th International Conference*, ICSMA 6 (ed. R.C. Gifkins), 1181–1195. Oxford: Pergamon Press.

Hatherly, M. and Malin, A.S. (1979). Deformation of copper and low stacking-fault energy, copper base alloys. *Met. Technol.* 6: 308–319.

Hatherly, M. and Malin, A.S. (1984). Shear bands in deformed metals. *Scr. Metall.* 18: 449–454.

Havner, K.S. (1992). *Finite Plastic Deformation of Crystalline Solids*. Cambridge: Cambridge University Press.

Hill, R. (1966). Generalized constitutive relations for incremental deformation of metal crystals by multislip. *J. Mech. Phys. Solids* 14: 95–102.

Honeycombe, R.W.K. (1984). *The Plastic Deformation of Metals*, 2nd ed. London: Edward Arnold.

Jasieński, Z. and Piątkowski, A. (1993). Nature de bandes de cisaillement macroscopiques dans les monocrisaux de cuivre solicites en compression plane. *Arch. Metall.* 38: 279–301.

Jasieński, Z. and Piątkowski, A. (1988). Shear bands formation in copper single crystals during plane-strain compression. *Strength of Metals and Alloys, Proceedings of the 8th International Conference 1CSMA 8*. Tampere, Oxford: Pergamon Press.

Karnop, R. and Sachs, G. (1928). Festigkeitseigenschaften von Kristallen einer veredelbaren Aluminiumlegierung. *Z. Phys.* 49: 480–497.

Kocks, U.F., Hasegawa, T., and Scattergood, R.O. (1980). On the origin of cell walls and lattice misorientation during deformation. *Scr. Metall.* 14: 449–454.

Korbel. A. (1985). The real nature of shear bands—plastons? *Plastic Instability, Proceedings of the Int Syrup on Plastic Instability, Considere Memorial (1841–1914)*. Paris: Presses de l'Ecole Nationale des Punts et Chaussees, pp. 325–335.

Korbel, A. (1987). Structural and mechanical aspects of localized deformation in Al-Mg alloy. *Arch Metall.* 32: 377–392.

Korbel, A. (1990a). The mechanism of strain localization in metals. *Arch. Metall.* 35: 177–203.

Korbel, A. (1990b). The model of microshear banding in metals. *Scr. Metall.* 24: 1229–1231.

Korbel, A. (1992). Mechanical instability of metal structure – catastrophic flow in single and polycrystals. *International Symposium on Advanced Crystal Plasticity*, pp. 42–68. Kingston: Canadian Institute of Mining and Metallurgy.

Korbel, A. and Bochniak, W. (1995). The structure based design of metal forming operations. *J. Mater. Process. Technol.* 53: 229–236.

Korbel, A. and Martin, P. (1986). Microscopic versus macroscopic aspect of shear bands deformation. *Acta Metall.* 34: 1905–1909.

Korbel, A. and Martin, P. (1988). Microstructural events of macroscopic strain localization in prestrained tensile specimens. *Acta Metall.* 36: 2575–2593.

Korbel, A. and Szczerba, M. (1988). Selfinduced change of deformation path in Cu-Al single crystals. *Rev. Phys. Appl.* 23: 706–711.

Korbel, A., Embury, J.D., Hatherly, M. et al. (1986). Microstructural aspects of strain localization in Al-Mg alloys. *Acta Metall.* 34: 1999–2009.

Kuhlmann-Wilsdorf, D. (1976). Recent progress in understanding of pure metal and alloy hardening. *Proceedings Symposium On Work Hardening in Tension and Fatique*, Cincinati, Ohio (11 November 1975) (ed. A.W. Thompson), 1–43. New York: AIME.

Kuhlmann-Wilsdorf, D. (1985). Theory of work hardening 1934–1984. *Metall. Trans.* A16: 2091–2108.

Langford, G. and Cohen, M. (1975). Microstructural analysis of plastic instabilities in strengthened metals. *Metall. Trans.* 6A: 901–910.

Lisiecki, L.L., Nelson, D.O., and Asaro, R.J. (1982). Lattice rotations, necking and localized deformation in fcc single crystals. *Scr. Metall.* 16: 441–448.

Luft, A. (1991), Microstructural processes of plastic instabilities in strengthenedmetals. Progress in Materials Science (ed. J.W. Christian, P. Haasen and T.B.Massalski), Pergamon Press, Oxford, 35: 98–180.

Malin, A.S. and Hatherly, M. (1979). Microstructure of cold—rolled copper. *Met. Sci. 13*: 463–172.

Masima, M. and Sachs, G. (1928). Mechanische Eigenschaften von Messing kristallen. *Z. Angew. Phys.* 50: 161–186.

Meyers, M.A. (1994). *Dynamic Behavior of Materials*. New York: Wiley.

Nemat—Nasser S. (2004). *Micromechanics, Plasticity. A Treatise on Finite Deformation of Heterogeneous Inelastic Materials*. UK: Cambridge University Press.

Neuhäuser H. (1983). Slip—line formation and collective dislocation motion. In: *Dislocations in Solids* (ed. F.R.N. Nabarro), 319–440: Elsevier, North-Holland, Amsterdam.

Nguyen, H.V. and Nowacki, W.K. (1997). Dynamic simple shear of metal sheets. *Arch. Mech.* 49: 369–384.

Oliferuk, W. (1995). Bilans energii a ewolucja mikrostruktury podczas jednoosiowego rozciągania stali austenitycznej. *Energy balance vis-à-vis microstructure evolution in the course of uniaxial tenstion of austenitic steel* (in Polish). *Rudy i Metale* 40: 438–441.

Oliferuk, W., Świątnicki, W.A., and Grabski, M.W. (1995). Effect of the grain size on the rate of energy storage during the tensile deformation of an austenitic steel. *Mater. Sci. Eng., A* 197: 49–58.

Oliferuk, W., Korbel, A., and Grabski, M.W. (1996). Mode of deformation and the rate of energy storage during uniaxial tensile deformation of austenitic steel. *Mater. Sci. Eng.* A220: 121–128.

Oliferuk, W., Korbel, A., and Grabski, M.W. (1997). Slip behaviour and energy storage process during uniaxial tensile deformation of austenitic steel. *Mater. Sci. Eng.* A234–A236: 1122–1125.

Orlov, L.G. and Shitikova, G.F. (1981). Structure changes during neck formation under tension of silicon iron crystals. *Fiz. Met. Metalloved.* 52: 421–424.

Paul, H., Jasieński, Z., Piątkowski, A. et al. (1996). Crystallographic nature of shear bands in polycrystalline copper. *Arch. Metall.* 41: 337–353.

Pęcherski, R.B. (1983). Relation of microscopic observations to constitutive modelling for advanced deformations and fracture initiation of viscoplastic materials. *Arch. Mech.* 35: 257–277.

Pęcherski, R.B. (1985). Discussion of sufficient condition for plastic flow localization. *Eng. Fract. Mech.* 21: 767–779.

Pęcherski, R.B. (1991). Physical and theoretical aspects of large plastic deformations involving shear banding. In: *Finite Inelastic Deformations. Theory and Applications. Proceedings of IUTAM Symposium*, Hannover, Germany (ed. D. Besdo and E. Stein), 167–178. Springer-Verlag.

Pęcherski, R.B. (1992). Modelling of large plastic deformations based on the mechanism of micro-shear banding. Physical foundations and theoretical description in plane strain. *Arch. Mech.* 44: 563–584.

Pęcherski, R.B. (1993). Theoretical description of plastic flow accounting for micro-shear bands. *Arch. Metall.* 38: 205–219.

Perzyna, P. (1971). Thermodynamic theory of viscoplasticity. In: *Advances in Applied Mechanics*, 313–354. New York: Academic Press.

Perzyna, P. (2012). Multiscale modelling of the influence of anisotropy effects on fracture phenomena in inelastic solids. *Eng. Trans.* 60: 225–284.

Phillips, R. (2001). *Crystals, Defects and Microstructures. Modelling Across Scales*. Cambridge: Cambridge University Press.

Pieczyska, E. (1999). Thermoelastic effect in austenitic steel referred to its hardening. *J Theor. Appl. Mech.* 37: 349–368.

Price, R.J. and Kelly, A. (1964). Deformation of age-hardened aluminium alloy crystals—II. Fracture. *Acta Metall.* 12: 979–992.

Puttick, K.E. (1963). Necking and fracture in aluminium crystals. *Acta Metall.* II: 986–989.

Raniecki, B. (1984). Thermodynamic aspects of cyclic and monotone plasticity. In: *The Constitutive Law in Thermoplasticity* (ed. Th. Lehmann), 251–321. Wien, New York: Springer-Verlag.

Richert, M. and Korbel, A. (1995). The effect of strain localization on mechanical properties of Al 99,992 in the range of large deformations. *J. Mater. Process. Technol.* 53: 331–340.

Romanov, A.E. and Vladimirov, V.I. (1983). Disclinations in solids. *Phys. Status Solidi A* 78: 11–34.

Rovelli, C. (2011). *The First Scientist Anaximander and His Legacy.* Yardley, PA: Westholme Publishing.

Rubtsov, A.S. and Rybin, V.V. (1978). Structural features of plastic deformation at the stage of localized flow. *Phys. Met. Metall.* 44: 139–149.

Russo, L. (2004). *The Forgotten Revolution. How Science Was Born in 300 BC and Why It Had to Be Reborn.* Berlin, Heidelberg: Springer-Verlag.

Rybin, V.V. (1978). Physical model of the phenomena of loss of mechanical stability and necking. *Phys. Met. Metall.* 44: 149–157.

Sachs, G. and Weerts, J. (1930). Zugversuche an Gold—Silberkristallen. *Z. Angew. Phys.* 62: 473–493.

Saimoto, S., Hosford, W.F. Jr., and Backofen, W.A. (1965). Ductile fracture in copper single crystals. *Philos. Mag.* 12: 319–333.

Schmid, E. and Boas, W. (1935). *Kristallplastizität.* Berlin: Springer-Verlag, Plasticity of Crystals with Special Reference to Metals, London, 1968.

Sevillano, J.G., Van Houtte, P., and Aernoud, T.E. (1982). Large strain work hardening and texture. *Prog. Mater. Sci.* 25: 69–412.

Spitzig, W.A. (1981). Deformation behavior of nitrogenated Fe—Ti—Mn and Fe—Ti single crystals. *Acta Metall.* 29: 1359–1377.

Szczerba, M. and Korbel, A. (1987). Strain softening and instability of plastic flow in Cu—Al single crystals. *Acta Metall.* 35: 1129–1135.

Taylor, G.I. and Elam, C.F. (1925). The plastic extension and fracture of aluminium crystals. *Proc. R. Soc.* A108: 28–51.

Taylor, G.I. and Quinney, H. (1934). The latent energy remaining in a metal after cold working. *Proc. R. Soc.* A143: 307–326.

Vegazov, A.N., Likhachev, V.A., and Rybin, V.V. (1977). Characteristic elements of the dislocation structure in deformed polycrystalline molybdenum. *Phys. Met. Metall.* 42: 126–133.

Wilsdorf, H.G.W. (1980). Dislocation pattern leading to the initiation of ductile fracture in pure metals and alloys. In: *Proceedings 5th International Conference on Strength of Metals and Alloys,* Aachen (27–31 August 1979), vol. 2 (ed. P. Haasen, V. Gerold and G. Kostorz), 1329. Toronto, Oxford, New York, Sydney, Paris, Frankfurt: Pergamon Press.

Wright, T.W. and Ravichandran, G. (1997). Canonical aspects of adiabatic shear bands. *Int. J. Plast.* 13: 309–325.

Wróbel, M., Dymek, S., and Blicharski, M. (1990). Microstructure of rolled copper single crystals. *Arch. Mech.* 35: 245–258.

I

Wróbel, M., Dymek, S., Blicharski, M., and Driver, J. (1995). Microstructural changes due to rolling of austenitic stainless steel single crystals with initial orientation (110)[001] and (110)[110]. *Scr. Metall. Mater.* 32: 1985–1991.

Yang S. and Rey C. (1993). Analysis of deformation by shear banding: A two-dimensional postbifurcation model, MECAMAT'91 (ed. C. Teodosiu et al.), Balkema, Rotterdam, 229–237.

Yang, S. and Rey, C. (1994). Shear band postbifurcation in oriented copper single crystals. *Acta Metall.* 42: 2763–2774.

Yeung, W.Y. and Duggan, B.J. (1987). On the plastic strain carried by shear bands in cold-rolled a-brass. *Scr. Metall.* 21: 485–490.

3

Incorporation of Shear Banding Activity into the Model of Inelastic Deformations

This chapter contains the following results of the earlier author's studies:

a) micromechanical foundations of theoretical description of finite inelastic deformations accounting for the shear banding mechanism summarised as Observation 3.1 and Hypothesis 3.1
b) the proposition of a new model determining a contribution to the rate of inelastic deformation produced by the multilevel hierarchy of micro-shear bands activity.

3.1 Plastic Deformation of Metallic Solids vis-à-vis the Continuum Mechanics

Integrated studies of physics and mechanics of plastic deformation of metals require critical analysis of the averaging procedures and macroscopic description of multilevel shear banding activity within the continuum theory of inelastic materials. The previously discussed dislocation groups move in an avalanche-like way along with active slip systems: slip lines and slip bands form coarse slip bands. Then micro-shear bands initiate and build up through clusters of micro-shear bands to macroscopic shear bands, thereby contributing to the rate of inelastic deformation. As mentioned earlier, the phenomena indicate that the crystalline body subjected to plastic deformation is a complex, multilevel, hierarchically organized system. The problem of determining the smallest representative volume element (RVE) required to transfer the variables relevant for a constitutive description of the macroscopic behaviour of heterogeneous materials (Bishop and Hill 1951), where the concept of a 'unit cube' of material transpired. Also, the issues related to the averaging procedure were the subject of analysis. Then Hill (1956), Hill (1963, 1967, 1984),

Viscoplastic Flow in Solids Produced by Shear Banding, First Edition. Ryszard B. Pęcherski.
© 2022 John Wiley & Sons Ltd. Published 2022 by John Wiley & Sons Ltd.

as well as Havner (1973), devoted much attention to clarify the physical foundations of the RVE concept. A comprehensive discussion of research results and extensive literature on the subject is contained in monographs (Havner 1992; Nemat-Nasser and Hori 1993). A broader view related to heterogeneous materials such as composites or multiphase materials is presented in works by Hashin (1964) and Kröner (1986), and monographs of Sobczyk (1991) and Nemat-Nasser and Hori (1993), as well as by Nemat-Nasser (2004). These authors provide an in-depth discussion of the theoretical basis for statistical physics adopting the ergodic hypothesis, cf. Truesdell (1966), de Oliveira and Werlang (2007), and Gallavotti (2016). De Oliveira and Werlang (2007) presented an updated discussion on the ergodic hypothesis's physical and mathematical aspects in classical equilibrium statistical mechanics. Gallavotti (2016) shared a comprehensive historical thread of this fundamental concept.

In the classical theory of plasticity of crystalline materials, which deals with various aspects of the description of plastic deformations produced by slips in easy systems, averaging stress and strain measures on certain RVE was the subject of interest and comprehensive research of many authors. According to Havner (1973, 1974, 1992), the use of a continuum description of finite deformations of metals limits to a minimum level of observation where physical features of inelastic behaviour of material are represented. Suppose one observes the surface of the deformed polycrystal with a resolution of up to $1\,\mu m$. In that case, the permanent deformation of a grain of about $100\,\mu m$ average diameter is relatively smooth. At the same level of observation, called by Havner microscopic one, we can distinguish slip lines that appear on the surface of a permanently deformed crystal or individual grains of a polycrystalline sample. Then, according to Havner (1992), p. 34: 'In contrast, a submicroscopic observer resolving distances to 10^{-5}mm (the order of 100 atomic spacings) is aware of highly discontinuous displacements within crystals. The microscopic observer's slip lines appear to the submicroscopic observer as slip bands of order 10^{-4}mm thickness, containing numerous glide lamellae between which amounts of slip as great as 10^3 lattice spacings have occurred, as first reported by Heidenreich (1949); hence a continuum perspective at the second level would seem untenable'.

Then, the physical dimension of RVE cannot be smaller than $1\,\mu m$, i.e. $>10^3\,a$. In this way, the RVE is idealized as a 'material point' of the continuum (its neighbourhood, more precisely). The minimum physical dimension is the following, according to Hill (1956), p. 8: '. . . dimensions must be large compared with the thickness of the glide packets separating the active glide lamellae (generally of order 10^{-4}cm in many metals at ordinary temperatures). Thus, the linear dimension of the smallest crystal whose behaviour can legitimately be considered from the standpoint of the theory of plasticity is probably of order 10^{-3}cm', which gives an estimate of the physical dimension up to the order of 10 μm. At the average grain size (10÷100) μm, the use of continuum mechanics to describe metal deformation is, as discussed above, physically correct. Bearing in mind the discussed limitations of the minimum level of observation, we can characterize the kinematics of crystal deformation. One of the essential phenomenological aspects of metals' global plastic behaviour is the continuum movement against crystal structure. It means that the material 'lines' and 'planes' move and rotate relative to the average orientation of the crystal lattice in volumes of a typical dimension of the order of 10 μm, cf. Havner (1992). Therefore, in these considerations, the average physical size of the volume element was determined, which guarantees the inelastic metal deformation's constitutive description at the microscopic level. However, for a continuum description of the elastoplastic strain on a macroscopic scale, such a dimension should be much larger than a multiple of the average grain size. For example, Havner (1992) assumes a polycrystalline unit cube with a 1 mm side containing about 1000 grains. Then the theory describing kinematics and constitutive structure of finite elastoplastic deformations is correctly formulated. The transition from the microscopic level to the macroscopic scale can be precisely determined, as well, cf. the pivotal results of Hill (1956, 1967, 1972), and Havner (1973, 1974, 1992) as well as Mandel (1971) and Hill and Rice (1972).

This discussion is valid under the general assumption that the dominant mechanism of inelastic behaviour is crystallographic slip. The theory describing kinematics and constitutive structure of finite elastic–plastic deformations of crystalline solids finds firm foundation in such a case. The transition between the microscopic and macroscopic levels is well understood (cf. e.g. Hill (1972, 1984, 1985), Havner (1973, 1974, 1992) as well as and Mandel (1974, 1980), and Hill and Rice (1972), and further Petryk (1989) and Stolz (1990)). In particular, the macro-measures of stress, strain, and plastic work correspond to their micro-measures' volume averages. The works mentioned above show that specific structural features of the theoretical continuum description of inelastic deformations, as the normality rule or specific constitutive inequalities, are transmitted upwards through a hierarchy of observational levels unchanged.

3.2 Hypothesis on the Extension of the RVE Concept

The situation changes essentially when the new multiscale shear banding mechanism is responsible for the inelastic deformation, cf. the analysis carried out in Pęcherski (1992, 1997) and Pęcherski and Korbel (2002). The discussion and summary of the available results of experimental observations included in Chapter 2 lead to the following view.

Observation 3.1

Micro-shear bands spread over a distance of at least a few grain diameters, often forming packets (clusters) of width (10–100) μm. These packets are thin layers of material. Successive generations of elementary active micro-shear bands, with a thickness of 0.1 μm, propagate rapidly relative to the environment.

The above view shows that the contribution of micro-shear bands in inelastic deformation appears at different levels of observation in contrast to the crystallographic slips developing only at the microscopic scale. The following issues were raised in the case of plastic deformation governed by micro-shear banding:

a) discontinuity of the component of the microscopic velocity field is tangential to the propagation plane of the active micro-shear band system
b) dynamic micro-stress redistribution processes appear
c) possible changes in the geometry of the RVE under consideration.

It causes fundamental difficulty while treating both mechanisms' effects, attempting to average them and make a theoretical transition to the macroscopic level. The question arises about the scale of observation, which is necessary to gain an adequate continuum description. Also, it is essential to account simultaneously for the effect of shear bands' multilevel hierarchy and crystallographic slips on the macroscopic material behaviour. These requirements produce contradictory opposites. Increasing the scale of the physical dimension of the volume element, which ought to be described as the infinitesimal neighbourhood of the 'material point' in the continuum description simultaneously, causes the 'blurring' of the effects of shear bands. It also deprives the possibility of considering the impact of their dominant features on the material's mechanical properties.

An example is the shear band geometry preserved at all observation levels, significantly affecting the plastic strain-induced anisotropy. The problem remains open. Further research is needed to develop a precise combination of material theory at the microscopic level of plastic slips and the mesoscopic level of micro-shear banding with the macroscopic description of the viscoplastic flow. This issue was the subject of the studies in Pęcherski (1997, 1998a,b). The previous discussion of the physical nature of the micro-shear banding process and recent microscopic observations *in situ*, presented by Yang and Rey (1993, 1994), and in particular fruitful and vivid discussions with Collete Rey, support the following hypothesis extending of the generally accepted concept of RVE.

Hypothesis 3.1

For a continuum description of inelastic deformation on a macroscopic scale, we will assume RVE as a unit volume of polycrystalline with an average linear dimension L_0 of the order of 1 mm. To take into account the impact of the multilevel hierarchy of micro-shear bands, we allow the possibility of the existence in RVE of the singular discontinuity surface of order one, of the microscopic velocity field on which the tangential component of velocity experiences a jump, i.e. the perturbation of the microscopic displacement field Δ_{ms} is travelling at the speed V_s. The orientation of this surface is determined by the geometry of the active micro-shear bands' system.

The discussion above allows presenting a proposition of a phenomenological model in the next chapter, which describes the inelastic behaviour of material within the range of finite strains accounting for the most important influence, for macroscopic description, of shear banding effects.

The approach proposed in this way seems to be adequate to the statement contained in the monograph (Nemat-Nasser and Hori 1993), p. 15, which says:

> Perhaps one of the most vital decisions that the analyst makes is the definition of the RVE (representative volume element). An optimum choice would be one that includes the most dominant features that have first-order influence on the overall properties of interest and, at the same time, yields the simplest model.

3.3 Model of Shear Strain Rate Generated by Micro-shear Bands

As per Observation 3.1 and Hypothesis 3.1, let us consider a certain RVE containing an area where the massive development of micro-shear bands occurs. Figure 3.1a depicts the traces of successive micro-shear band packets (clusters) activity. The arrow indicates the direction of the expansion of the macroscopic shear band zone. This zone displays two consecutive

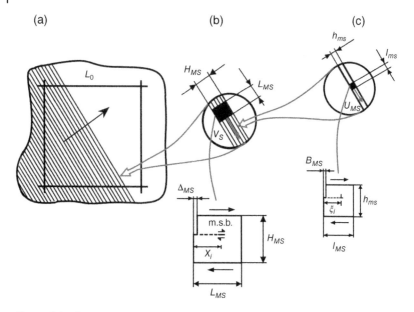

Figure 3.1 Schematic illustration of the multilevel hierarchy of micro-shear bands: (a) RVE image of the average linear dimension L_0 with shear band zone expanding in the direction of the arrow; (b) 'magnification' of the micro-shear bands' cluster of the width H_{ms} and the length L_{ms} of the working area (active zone), (c) 'zoom in' of the micro-shear band active area of the width h_{ms} and the length l_{ms} with a schematic view of the dislocation mechanism of shear producing the local perturbation B_{ms} of the microscopic displacement field. *Source:* Copyright by Aleksandra Manecka – Padaż.

'magnifications', Figure 3.1b,c, with a schematic view of underlying shearing mechanisms. Figure 3.1b shows the zoom micro-shear bands with an active area of thickness H_{MS} and length L_{MS}. The passage of many active micro-shear bands produces the local perturbation Δ_{MS} of the microscopic field of displacements propagating with the velocity V_S as a distortion wave. The second stage of 'magnification' is illustrated in Figure 3.1c. It represents the zoom-in of an active area of a single micro-shear band with h_{ms} thickness and l_{ms} length, causing local perturbation of the B_{ms} microscopic displacement field moving on the forehead of the micro-shear band as a distortion wave at the v_{ms} speed. The schematic view of the edge dislocation transition is also displayed. According to the known concept of Gilman (1960), we have:

$$\delta_i = \frac{\xi_i b}{l_{ms}}, \quad B_{ms} = \sum_i^n \delta_i = \frac{b}{l_{ms}} \sum_i^n \xi_i \qquad (3.1)$$

where δ_i denotes the displacement of the "i"th dislocation moving in the active area, while B_{ms} corresponds to the movement of the upper part of the working zone produced by n dislocations. The resulting shear strain γ reads:

$$\gamma = \frac{B_{ms}}{h_{ms}} = \frac{bn}{l_{ms}h_{ms}}\bar{\xi}, \; \bar{\xi} = \frac{\sum_i^n \xi_i}{n} \qquad (3.2)$$

where $\bar{\xi}$ denotes the mean distance the edge dislocations are travelling in the active region in Figure 3.1c. Assuming that the sizes of $\bar{\xi}$ and n may change during the duration of the microscopic plastic shear process, which we denote as the τ variable, we can calculate the corresponding shear strain rate

$$\frac{d\gamma}{d\tau} = \frac{b}{l_{ms}h_{ms}}\left(nv_d + \bar{\xi}\frac{dn}{d\tau}\right) \qquad (3.3)$$

where

$$v_d = \frac{d\bar{\xi}}{d\tau} \qquad (3.4)$$

is the average dislocation speed. Thus, the shear strain rate generated by a single micro-shear band is related to the speed v_{ms} associated with the propagation of the active zone, moving with its forehead:

$$\frac{d\gamma}{d\tau} = \frac{v_{ms}}{h_{ms}}, \; v_{ms} = \frac{b}{l_{ms}}\left(nv_d + \bar{\xi}\frac{dn}{d\tau}\right). \qquad (3.5)$$

As noted in Section 2.1.1, in the case of micro-shear band propagation through many grains, the so-called difficult slip systems characterizing with the high values of critical resolved shear stress may activate. It generates new dislocations in the active zone, and therefore the presence of the second component in (3.5) is theoretically fully justified. However, it is difficult to experimentally determine the mutual interaction of generation and dislocation movement mechanisms and their contribution to the shear strain rate. It remains the subject of further studies with the application of molecular dynamics (MD) simulations. Therefore, as a first approximation, one can assume that plastic shear strain is generated only by the movement of a certain averaged number of dislocations that do not change during the propagation of the active zone. Then the relation (3.5) simplifies to the known Orowan formula:

$$\frac{d\gamma}{d\tau} = b\rho\upsilon_d, \, \rho = \frac{n}{l_{ms}h_{ms}}, \quad (3.6)$$

where ρ denotes dislocation density.

Consider a certain number of active micro-shear bands N_{MS} of similar orientation and produced within a time interval $\Delta\tau = \tau_f - \tau_i$. The macroscopic description of plastic flow is so small that it is a 'short while'. Strictly speaking, it represents an infinitely small increase δt by a 'timelike parameter'. A two-level time scale concept appears here Pęcherski (1997, 1998a,b) and (Pęcherski and Korbel (2002). We note a similarity to the representative time increment (RTI) idea introduced by Petryk (1995). In our considerations, the RTI corresponds to the interval $\Delta\tau$. The cluster of micro-shear bands acting simultaneously (averaged over $\Delta\tau$) understood, in this way, and depicted in Figure 3.1b is responsible for microscopic shear strain γ_{ms}

$$\gamma_{ms} = \frac{\Delta_{MS}}{H_{MS}}, \, \Delta_{MS} = \frac{\overline{B}_{ms}N_{MS}}{L_{MS}}\overline{x}_{MS} \quad (3.7)$$

where $\overline{B}_{ms}$ is the total displacement resulting from a single micro-shear band passage

$$\overline{B}_{ms} = h_{ms}\overline{\gamma} = \int_{\tau_i}^{\tau_f}\upsilon_{ms}d\tau, \quad (3.8)$$

while

$$\overline{x}_{MS} = \frac{\sum_i^k x_i}{k}, \, k = N_{MS} \quad (3.9)$$

denotes the mean path passed by N_{MS} of micro-shear bands propagating in the active zone of the cluster. The length of the active zone L_{MS} is determined as a mean operating range of micro-shear bands propagating with the average velocity υ_{ms} within the time interval $\Delta\tau = \tau_f - \tau_i$:

$$L_{MS} = \upsilon_{ms}\Delta\tau. \quad (3.10)$$

Allowing the possibility of changing x and N during the propagation of the active cluster zone, we have, based on (3.7):

$$\dot{\gamma}_{ms} = \frac{\overline{B}_{ms}}{L_{MS}H_{MS}}\left(N_{MS}\dot{\overline{x}}_{MS} + \overline{x}_{MS}\dot{N}_{MS}\right), \quad (3.11)$$

where the dot means differentiation over the 'timelike parameter' t. Let us observe that the velocity $\dot{\bar{x}}_{MS}$ can be identified with the speed of propagation of the head of the active zone of a single micro-shear band, $\dot{\bar{x}}_{MS} = v_{ms}$, under the simplifying assumption that v_{ms} is of the same value for each of the micro-shear bands traversing the active zone of the cluster. Then, the propagation velocity V_S of the working area of the cluster of N_{MS} micro-shear bands takes the form:

$$V_S = \frac{\bar{B}_{ms}}{L_{MS}}\left(N_{MS}\dot{\bar{x}}_{MS} + \bar{x}_{MS}\dot{N}_{MS}\right) \tag{3.12}$$

and the corresponding strain rate reads:

$$\dot{\gamma}_{ms} = \frac{V_S}{H_{MS}}. \tag{3.13}$$

The relation (3.12) shows that the additional evolution equation for the number of micro-shear bands N_{MS} is necessary. It creates an open problem requiring further theoretical and experimental studies on developing micro-shear bands in a single cluster's active working area. Suppose for simplicity, however, that the number of micro-shear bands during the propagation of a single cluster active functional area is approximately constant, then the relation (3.12) simplifies as follows

$$V_S = \frac{\bar{B}_{ms}}{L_{MS}}N_{MS}v_{ms}, \tag{3.14}$$

and (3.13) takes the Orowan-like form

$$\dot{\gamma}_{ms} = \bar{B}_{ms}\rho_{ms}v_{ms}, \rho_{ms} = \frac{N_{MS}}{L_{MS}H_{MS}}, \tag{3.15}$$

where ρ_{ms} denotes the density of micro-shear bands operating in the active working area of the cluster. It is the number of active micro-shear bands performed within the active zone of the cluster. According to the estimation foreseen in Pęcherski (1998a), for N_{MS} assumed to be of order 100 and H_{MS} and L_{MS} being about 100 μm, the density ρ_{ms} 10^{10} (m^{-2}) is reached. The experimental observations and particular analogy with martensitic transformations led to the conjecture that micro-shear bands propagate with the velocity v_{ms} constant magnitude bounded by the elastic shear wave speed c_s, i.e. the velocity of sound in the considered metal or alloy (Pęcherski 1997). Thus, the following relation appears justified

$$\upsilon_{ms} = \eta c_s = \eta \left(\frac{\mu}{\rho} \right)^{1/2}, \eta \in (0,1), \tag{3.16}$$

where η relates to the effect of dissipation related to nucleation and movement of dislocations in the activated system, the value of η could be attempted to assess from $(3.5)_2$. However, it isn't easy to evaluate consistently from experimental data of all the structural parameters occurring in relation (3.5). Therefore, the direct measurements seem to be more appropriate. For instance, shear band speed measurements in a $C - 300$ steel (high-strength maraging steel) were reported (Zhou et al. 1996). The highest value of speed observed is close to $1200 \, \mathrm{ms}^{-1}$, which means approximately 40% of the shear wave speed c_s of the specimen material. It provides the estimate $\eta \cong 0.40$. Let us note that the investigated shear bands correspond to the clusters of micro-shear bands. According to (3.14), the speed of a single micro-shear band should be much higher, which produced also a higher value of η. The complementary assessment of the amounts of structural parameters:

- the total displacement resulting from a single micro-shear band passage $\bar{B}_{ms}$
- the length of the active zone of the cluster L_{MS}
- the number of micro-shear bands propagating in the active zone of the cluster N_{MS}

characterizing the investigated shear bands is necessary to obtain a more exact specification of η.

References

Bishop, J.F.W. and Hill, R. (1951). A theoretical derivation of the plastic properties of a polycrystalline face-centered metal. *Philos. Mag.* 42: 1298–1307.

De Oliveira, C.R. and Werlang, T. (2007). Ergodic hypothesis in classical statistical mechanics (Hipothese ergodica em mecanica estatistica classica). *Rev. Brasil. Ensino de Fisica* 29: 189–201.

Gallavotti, G. (2016). Ergodicity: a historical perspective. Equilibrium and Nonequilibrium. *The European Physical Journal H* 41: 181–259.

Gilman, J.J. (1960). Physical nature of plastic flow and fracture, plasticity. In: *Proceedings of the 2nd Symposium on Naval Structural Mechanics* (ed. E.H. Lee and P.S. Symonds), 43–99. Pergamon Press.

Hashin, Z. (1964). Theory of mechanical behavior of heterogeneous media. *Appl. Mech. Rev.* 17: 1–9.

Havner, K.S. (1973). On the mechanics of crystalline solids. *J. Mech. Phys. Solids* 21: 383–394.

Havner, K.S. (1974). Aspects of theoretical plasticity at finite deformation and large pressure. *ZAMP* 25: 765–781.

Havner, K.S. (1992). *Finite Plastic Deformation of Crystalline Solids*. Cambridge: Cambridge University Press.

Heidenreich, R.D. (1949). Structure of slip bands and cold worked metal. *Trans. ASM A* 41: 57–64.

Hill, R. (1956). The mechanics of quasi-static plastic deformation in metals. In: *Surveys in Mechanics* (ed. C.K. Batchelor and R.M. Davies), 7–31. Cambridge: Cambridge University Press.

Hill, R. (1963). Elastic properties of reinforced solids: Some theoretical principles. *J. Mech. Phys. Solids* 11: 357–372.

Hill, R. (1967). The essential structure of constitutive laws for metal composites and polycrystals. *J. Mech. Phys. Solids* 15: 779–795.

Hill, R. (1972). On constitutive macro-variables for heterogeneous solids at finite train. *Proc. R. Soc. London, Ser. A* 326: 131–147.

Hill, R. (1984). On the micro-to-macro transition in constitutive analyses of elastoplastic response at finite strain. *Math. Proc. Cambridge Philos. Soc.* 95: 481–494.

Hill, R. (1985). On macroscopic effects of heterogeneity in elastoplastic media at finite strain. *Math. Proc. Cambridge Philos. Soc.* 98: 579–590.

Hill, R. and Rice, J.R. (1972). Constitutive analysis of elastic-plastic crystals at arbitrary strain. *J. Mech. Phys. Solids* 20: 401–413.

Kröner, E. (1986). The statistical basis of polycrystal plasticity. In: *Large Deformations of Solids Physical Basis and Mathematical Modelling* (ed. J. Gittus, J. Zarka and S. Nemat-Nasser), 27–40. London and New York: Elsevier.

Mandel, J. (1971). *Plasticitè classique et viscoplasticitè. C.I.S.M.* Udine: Springer-Verlag.

Mandel, J. (1974). Thermodynamics and plasticity. In: *Foundations of Continuum Thermodynamics* (ed. J.D. Domingos, M.N.R. Nina and J.H. Whitelaw), 283–304. London: McMillan.

Mandel, J. (1980). Mècanique des Solides anèlastiques — Gènèralisation dans R9 de la règle du potentiel plastique pour un èlèment polycrystalline. *C.R. Acad. Sci. Paris* 290B: 481–484.

Nemat-Nasser, S. (2004). *Micromechanics, Plasticity. A Treatise on Finite Deformation of Heterogeneous Inelastic Materials*. Cambridge University Press.

Nemat-Nasser, S. and Hori, M. (1993). *Micromechanics, Overall Properties of Heterogeneous Materials*, second revised edition (1999). Amsterdam: Elsevier.

Pęcherski, R.B. (1992). Modelling of large plastic deformations based on the mechanism of micro-shear banding. Physical foundations and theoretical description in plane strain. *Arch. Mech.* 44: 563–584.

Pęcherski, R.B. (1997). Macroscopic measure of the rate of deformation produced by micro-shear banding. *Arch. Mech.* 49: 385–401.

Pęcherski, R.B. (1998a). Macroscopic effects of micro-shear banding in plasticity of metals. *Acta Mech.* 131: 203–224.

Pęcherski, R.B. (1998b). Macromechanical description of micro—shear banding. In: *Proceedings of the McNU '97 Symposium on Damage Mechanics in Engineering Materials*, Studies in Applied Mechanics, June 28 — July 2, 1997, Evanston, vol. 46 (ed. G.Z. Voyiadjis, J.W. Ju and J.-L. Chaboche), 203–222. New York: Elsevier.

Pęcherski, R.B. and Korbel, K. (2002). Plastic strain in metals by shear banding. I. Constitutive description for simulation of metal shaping operations. *Arch. Mech.* 54: 603–620.

Petryk, H. (1989). On constitutive inequalities and bifurcation in elastic-plastic solids with a yield-surface vertex. *J. Mech. Phys. Solids* 37: 265–291.

Petryk, H. (1995). Thermodynamic stability of equilibrium in plasticity. *J. Non-Equlib. Thermodyn.* 20: 132–149.

Sobczyk, K. (1991). *Stochastic Differential Equations With Applications to Physics and Engineering*. Dordrecht, Boston, MA, London: Kluver Academic Publisher.

Stolz, C. (1990). On relationship between micro and macro scales for particular cases of nonlinear behaviour of heterogeneous media. In: *Proceedings of IUTAM/ICM Symposium on Yielding, Damage and Failure of Anisotropic Solids* (ed. J.-P. Boehler), 617–628. London: Mechanical Engineering Publications.

Truesdell, C. (1966). *Six Lectures on Modern Natural Philosophy*. Berlin, Heidelberg: Springer-Verlag.

Yang, S. and Rey, C. (1993). Analysis of deformation by shear banding. A two-dimensional post-bifurcation model. In: *MECAMAT'91* (ed. C. Teodosiu), 229–237. Rotterdam: Balkema.

Yang, S. and Rey, C. (1994). Shear band postbifurcation in oriented copper single crystals. *Acta Metall.* 42: 2763–2774.

Zhou, M., Rosakis, A.J., and Ravichandran, G. (1996). Dynamically propagating shear bands in impact loaded pre-notched plates. I. Experimental investigations of temperature signatures and propagation speed. *J. Mech. Phys. Solids* 44: 981–1006.

4

Basics of Rational Mechanics of Materials

According to the previous chapter, Section 3.1 on the physical motivation for the theoretical description of inelastic deformations produced by shear banding, crystalline material's representative volume element (RVE) is the configuration of a body element idealised as a particle. The particle configuration at a particular instant and place in the three-dimensional Euclidean point space ε_3 is precisely the material point with its infinitesimal neighbourhood. As seen in Section 4.2.1, the particle becomes a carrier of the multiscale shearing producing the viscoplastic flow. Then, a novel concept transpires of the particle transmitting the information about the multilevel shear banding in the crystalline body element.

4.1 A Recollection of Rational Continuum Mechanics

In the face of ever-changing research trends, one might imagine that the somewhat distant past could affect the proper understanding of the subject for the younger generation of potential readers. Therefore, I am devoting a small historical outline to rational mechanics and the main characters associated with the approach to the matter.

The innovative ideas, enriched with the rational spirit of the great scholars such as James Bernoulli (1654–1705), Leonhard Euler (1707–1783), Joseph Louis Lagrange (1736–1813), and Augustin Louis Cauchy (1789–1857), permeated the European science as a subtle golden braid of rational mechanics, recalling the insightful metaphor of Hofstadter (1979). The sincere supporter of the age of enlightenment, Clifford Ambrose Truesdell (1919–2000), started to weave this thread in the 1950s. Together with Walter Noll (1925–2017) and their keen coworkers, they

Viscoplastic Flow in Solids Produced by Shear Banding, First Edition. Ryszard B. Pęcherski.
© 2022 John Wiley & Sons Ltd. Published 2022 by John Wiley & Sons Ltd.

ensured rational approach to the theory of materials and the related mathematical infrastructure became the tenets of modern research in mechanics worldwide. It is an effect of unprecedented growth of popularity of the work of Truesdell and Toupin (1960) and the successive editions of the pivotal monograph by Truesdell and Noll (1992), see the detailed information about numerous reprints, new editions, and translation into Chinese by Bertram (2014, p. 121). Also worth mentioning is the book of Truesdell (1977), cf. also the essays and papers dedicated to the memory of Clifford Truesdell edited by Man and Fosdick (2004), and the scientific biography of Walter Noll by Ignatieff (1996). According to the quoted biographical data, it is evident that both pioneers of rational mechanics and rational thermodynamics cf. Truesdell (1984) made a significant impact on the development of this field during the second half of the last century. Numerous lectures, seminars, and cooperation with the leading scientific centres and research groups worldwide caused widespread theoretical foundations of rational thermomechanics of materials.

A significant example is a little book by Truesdell (1966) presenting crucial modern natural philosophy. The author discusses the following contents vividly:

I)   Rational mechanics of materials
II) Polar and oriented media
III) Thermodynamics of viscoelasticity
IV) Electrified materials
V)   The ergodic problem in classical statistical mechanics
VI) Method and taste in natural philosophy.

In the first lecture, the three basic concepts became the primitive elements of rational thermomechanics of materials (Truesdell 1966, p. 2):

1) The concept of body $\mathcal{B}$ is a three-dimensional smooth manifold. The body's elements are particles **X**. Two important statements sum up the observations common to all gross bodies.
   a) Every sufficiently smooth portion of the body is a body, and a measure called mass is distributed smoothly over the body.
   b) A body is never accessible to direct observations, and one encounters it only at particular times and particular places. These times and places form the space–time manifold – Euclidean and three-dimensional.
2) Euclidean space–time consists of points x and is endowed in the Euclidean metric structure, independent of the time t. A moving body B

is mapping smoothly onto the regions of Euclidean space–body configurations in continuum mechanics.

3) A system of forces. In continuum mechanics, contact forces express material's reaction upon the material in a body's configuration:

$$t_n = \sigma n,$$

where t_n denotes the stress vector acting upon the boundary with outer normal n and σ is the Cauchy stress tensor.

Application of known conservation laws for force and torque, i.e. for linear momentum and moment of momentum, leads to Cauchy's laws of motion:

$$div\sigma + \rho b = \rho \ddot{x}, \sigma = \sigma^T,$$

where ρ is the mass density, b denotes the body force, $\ddot{x}$ is the acceleration, and superscript T means transposition.

Furthermore, the main principles of rational continuum mechanics are explained very clearly, Truesdell (1966, p. 6):

1) *Determinism.* The stress in a body is determined by the motion of the body has undergone.

2) *Local action.* The motion outside an arbitrary small neighbourhood of a particle may be disregarded in determining the stress at that particle.

3) *Material indifference.* Any two observers of a motion of a body find the same stress.

It is worth mentioning that the Polish translation of this little book appeared within three years – a real *signat tempora (a sign of the times).*

As a Gdańsk Polytechnic student at the beginning of 1970s, fascinated by the Polish translation of Truesdell (1969) and the mathematical foundations, as Walter Noll said: mathematical infrastructure, of rational continuum mechanics and thermodynamics, I got from Walter Noll a kind answer to my request letter, together with a bulky packet of his current papers. I also remember my professors' recollections of Clifford Truesdell and Walter Noll's visits in the late sixties and early seventies of the last century in my home Institute of Fundamental Technological Research of the Polish Academy of Sciences in Warsaw, to recall two of them: Henryk Zorski (1927–2003) – head of the Department of Theory of Continuous Media and Antoni Sawczuk (1927–1984) – head of the Department of Mechanics of Continuous Media. They organised invited lectures and seminars of these eminent scholars. An outcome of the increased interest in the rational approach to continuum mechanics and its mathematical foundations led to Jan Rychlewski (1934–2011) lecture courses on the tensor theory applications in nonlinear continuum mechanics (Rychlewski 1969), cf Ostrowska-Maciejewska (2020a).

An excellent example of the acquired new teaching standards on continuum mechanics is Smith's (1993) book *An Introduction to Continuum Mechanics – after Truesdell and Noll*. Walter Noll's influence upon research into the foundations of mechanics and thermodynamics of materials is straight, acknowledged everywhere. Another example of keeping the standards of rational mechanics alive is the monograph on the phenomenological mechanics of continuous media published recently in Polish (Ostrowska-Maciejewska 2020b). The specific feature of this work is an in-depth approach to elastic anisotropy and the theory of elastic limit states of anisotropic solids, initiated in Olszak and Ostrowska-Maciejewska (1985), developing novel Rychlewski's concepts of anisotropic solids and proper elastic states. These are, namely, the theorems on the nonlinear (spectral)-invariant decomposition (Rychlewski 1995a,b), the linear (harmonic)-invariant decomposition (Rychlewski 2000, 2001) of Hooke's tensors, and the limit quadratic condition for any linear elastic anisotropy possessing unique energy interpretation (Rychlewski 1984). A more detailed discussion of this subject is presented in Chapter 7.

4.2 The Rational Theory of Materials – Epilogue

Encouraged by the results of Noll (1972) on the theory of materials, Perzyna and Kosiński (1973) published an alternative mathematical theory of materials. They contributed their state-arising concept due to a novel *method of preparation* idea. Later on, they further extended the material theory for work-hardening plasticity and viscoplasticity within the theoretical framework of the *method of preparation* (Frischmuth et al. 1986).

Bertram (2014) published a critical review of the material theory's history and an in-depth discussion of possible further development perspectives. He presented two approaches that existed until the 1970s: history functional with fading memory and internal variables with the evolution equations. He found some theoretical and practical applications for viscous fluids, viscometric flows, and viscoelastic materials. The contemporary technological challenges and resulting academic demands made Walter Noll propose a *New Mathematical Theory of Simple Materials* (Noll 1972). Noll and his coworkers constantly developed the rigorous mathematical infrastructure that gained generality and mathematical elegance. However, as Bertram (2014, p. 119) stated: 'The gain in generality was, however, accompanied by a loss of simplicity. Consequently, this theory has been used only by very few groups within the scientific community'. Therefore, since then, most publications on material modelling have become more practical and pragmatic without claiming mathematical purity and generality.

Notwithstanding, Bertram (2014) expressed the need and well-founded requirement for a general thermodynamic theory accounting for the elas-tic–plastic and viscoplastic materials accounting for the development of damage and new functional materials more critical from an industrial point of view. He also noted the authors' above-mentioned attempt (Frischmuth et al. 1986) to develop the material theory with an additional concept of the state equipped with the *method of preparation*. It was one of the proposals put to the test of time. More recent extensive works (Perzyna 2010, 2012) accounted for particular applications of the theory with the *method of preparation* in the field of nanocrystalline metals and shear banding effects description. One can consider these pivotal works as the theoretical supplements that form the proper mathematical basis for more theoretically demanding readers instead of directly and intuitively displaying the crucial ideas of viscoplastic flow produced by shear banding. They are generally available for free access on the quarterly Engineering Transactions website. The *method of preparation* that was also used by Piotr Perzyna (1931–2013) in monograph (Perzyna 1978), consistently dis-cussed, among others, different classes of mathematical material struc-tures: material structure with internal parameters, the material structure with fading memory, the ones of differential type, and the rate type, and of mixed type: the one with fading memory and internal parameters or the material structure with internal parameters and of differential type or the rate type. Then an extensive chapter on elastic–viscoplastic material and elastic–plastic, as well as viscoelastic materials, followed. Furthermore, the treatise by Perzyna (1978) distinguishes a beneficial and considerable number of references – 1165 positions.

Krzysztof Wilmański (1940–2012) published the pivotal works, consid-ered as one of the cornerstones of the recent approach to the theory of mate-rials (Wilmański 1998, 2008) that are worthy of careful studies and recommendable to the more demanding reader. In particular, to develop the deeper basis of thermomechanical processes analysis, the explicit goal of the present book attempts mainly to focus on isothermal shear banding pro-cesses. As stated in the Preface of Wilmański (1998): '*This book has a differ-ent form from that usually found in books on continuum mechanics and continuum thermodynamics. The presentation of the formal structure of con-tinuum thermodynamics is not always as rigorous as a mathematician might anticipate . . .'.* It makes a visible advantage because the reader can understand the core of the subject's foundations. The monograph (Wilmański 2008) provides the classical continuum mechanics and thermo-mechanics presentation completed with detailed applications and examples Albers and Wilmański (2015). It is worthwhile to note that Wilmański (2008) is underlining a historical thread extensively. The eminent scholars who contributed distinctly to the development of nonlinear mechanics and rational mechanics of materials are: Ronald S. Rivlin (1915–2005), Jerald L. Ericksen (1924–2021), and James G. Oldroyd (1921–1982), as well as Leonid I. Sedov (1907–1999).

4.2.1 The Concept of the Deformable Body

Based on Smith's (1993) presentation of mathematical foundations and the basic concepts of rational continuum mechanics, the discussed subject will be outlined to extend the above-mentioned Truesdell's initial announcements. A body $\mathcal{B}$ is a three-dimensional smooth differentiable manifold being a set of uncountable elements called particles. A body is an abstract concept never accessible directly to observation. We encounter it only at particular times and in specific places as configurations in a three-dimensional real Euclidean point space ε_3 with associated vector or translation space E_3. Only in some placement (configuration) is a body ever likely observed (Truesdell 1966, 1977). The particle corresponds to the smallest part of the body, amenable to a phenomenological description. The body element placement relates to the material point – the representative volume element configuration of crystalline material; see the discussion in Section 3.1. The particles that are body elements interact with each other and are under the influence of external agents. As it turned out, the particle understood in this way was a carrier of viscoplastic flow's multi-scale shear banding mechanism. It is a novel concept of the particle endowed with the transfer of information on a multilevel hierarchy of micro-shear bands developed in the body element of crystalline material, Figure 3.1. Such a concept resembles a particle endowed with fading memory in rheological materials cf., e.g. Coleman and Noll (1961) and Perzyna (1978). Gerhard Goldbeck critically emphasises the difficulties and short-comings of applying a traditional direct multiscale integration scheme often called less precisely as 'micro-macro transition'. He strongly advocates and motivates a novel competitive approach based on multilevel information workflow in cyber manufacturing processes. The book's fore-word presents this idea in Meunier (2012) by collecting papers on molecular modelling and simulation techniques in material science computational methods.

4.2.2 The Motion of the Body

The motion of body $\mathcal{B}$ is presented in a framing. The framing or an observer is a triple $(\varepsilon_3, \mathcal{T}, \Phi)$ consisting of physical Euclidean point space ε_3, one-dimensional space $\mathcal{T}$ representing physical time, and a function Φ that maps the event world $\mathcal{W}$ homeomorphically onto the product space $\varepsilon_3 \times \mathcal{T}$. It means that the mapping is one-to-one and both Φ and Φ^{-1} are continuous. According to Noll (1973) and Truesdell (1977), the event world $\mathcal{W}$ is a topological space of events that an observer is mapping into his own physical space.The body points, e.g. X, Y, Z, correspond to material objects' idealisations. The motion χ_t at time $t \in \mathcal{T}$ takes body points X of $\mathcal{B}$ onto places $\mathbf{x}$ in ε_3:

$$\mathbf{x} = \chi_t(X) = \chi(X, t) \text{ for } X \in \mathcal{B}, \tag{4.1}$$

so that the body point X will occupy the place $\mathbf{x} = \chi_t(X)$ at time t as a result of the motion as observed by Φ. The placement of the body point X in ε_3 is the material point $\mathbf{x}$ belonging to the configuration of $\mathcal{B}$ at time t in the framing Φ. A motion of $\mathcal{B}$ is intrinsic, independent of the observer (framing). In Smith (1993), the other framings Φ^* may determine the motion, leading to the discussion of frame indifference, cf. also, e.g. Truesdell (1966). Specific fields of classical mechanics as force and, in particular, stress fields are frame indifferent. Many other fields, however, as the velocity or acceleration fields of a motion, are not.

In the framing Φ, the velocity $\mathbf{v}$, and acceleration $\mathbf{a}$ of the body point X at time t in the motion χ we define as

$$\mathbf{v} \equiv \dot{\chi} \equiv \frac{d\chi(X, t)}{dt} = \lim_{\Delta t \to 0} \frac{\chi(X, t + \Delta t) - \chi(X, t)}{\Delta t} \tag{4.2}$$

and

$$\mathbf{a} \equiv \ddot{\chi} \equiv \frac{d^2\chi(X, t)}{dt^2} = \frac{d\mathbf{v}}{dt}, \tag{4.3}$$

provided the motion is sufficiently smooth in t. The body $\mathcal{B}$ is a smooth manifold and, in particular, mappings as χ_t are as smooth as may be required. As we shall observe in the next chapter, the smoothness requirements may be weaker at points of discontinuity in studying, in our case, a singular surface of strong discontinuity of order one. The body is also a measure space with nonnegative-valued Borel measure giving the mass m on all Borel subsets $\mathcal{P}$ of $\mathcal{B}$ providing the body's mass distribution. Hence the mass $m(\mathcal{P})$ of any regular part $\mathcal{P}$ of $\mathcal{B}$ reads

$$\text{Mass of } \mathcal{P} = m(\mathcal{P}) = \int dm \text{ for any sub-body } \mathcal{P} \text{ of } \mathcal{B}. \tag{4.4}$$

4.2.3 The Deformation of a Body

A motion of a body $\mathscr{B}$ is a one-parameter family of maps (smooth diffeomorphisms – as smooth as may be required) $\chi = \chi_t = \chi\,(.,t)$, where the map χ_t is the placement (configuration) of $\mathscr{B}$ at time t during the considered motion. Let us note that the smoothness requirements may be weaker at discontinuity points, lines, or surfaces, while shock-type motions or singular surfaces of order one are studied. It will be the subject of analysis in the sequel.

One of the configurations can be a reference placement κ and the transplacement or deformation vis-à-vis the reference configuration reads:

$$\chi_\kappa := \chi_t \, \kappa^{-1} \tag{4.5}$$

According to the permanence of matter principle, finite positive volume regions should not deform into zero or infinite volume parts. Hence the maps χ_t and κ must be one-to-one mappings with the corresponding inverses χ_t^{-1} and κ^{-1}. Then the related deformation χ_κ is the one-parameter family of maps given by the composition of χ_t with κ^{-1},

$$\chi_\kappa \equiv \chi_\kappa\big(X,t\big) := \chi_t \circ \kappa^{-1}. \tag{4.6}$$

Usually, the reference places one denotes with upper-case bold letters:

$$\mathbf{X} = \kappa\big(X\big), \mathbf{Y} = \kappa\big(Y\big), \text{etc}, \tag{4.7}$$

for body points X, Y, in $\mathscr{B}$, whereas lower-case bold letters denote the actual places occupied during the motion:

$$\mathbf{x} = \chi\big(X,t\big), \mathbf{y} = \chi\big(Y,t\big), \text{etc}. \tag{4.8}$$

Then the transplacement χ_κ is mapping the reference places $\mathbf{X}$ onto actual places $\mathbf{x}$ according to:

$$\mathbf{x} = \chi_\kappa\big(X,t\big), \tag{4.9}$$

in the referential description of a motion. The transplacement χ_κ is also known as the deformation.

4.2.4 The Deformation Gradient

The local deformation tensor or the local deformation gradient $\mathbf{F} = \mathbf{F}(\mathbf{X}, t)$ defines for reference places $\mathbf{X} = \kappa(X)$ of the body point X in $\mathscr{B}$, as the spatial gradient of the deformation,

$$\mathbf{F}\big(\mathbf{X},t\big) = \nabla \chi_\kappa\big(\mathbf{X},t\big) \text{ for fixed } t. \tag{4.10}$$

The symbol of gradient $\nabla \chi_\kappa$ is expressed equivalently often using the derivative Grad χ_κ as, e.g. in Wilmański (1988). According to the detailed discussion in Smith (1993, p. 41), **F** denotes a second-order tensor being a linear operator determined on three-dimensional vector space E_3. One assumes that the mapping $\chi_\kappa(\cdot, t)$ is invertible, which means that each place **x** can be occupied only by one material point **X**. Consequently, this requires

$$det\ \mathbf{F} \neq 0. \tag{4.11}$$

Instead of developing the formal structure of continuum mechanics, let us follow Wilmański's teaching path (Wilmański 1998, 2008), who showed 'the core of the deformation gradient in a pictorial, demonstrative way'.

In Figure 4.1, we illustrate the meaning of the deformation gradient **F** as a linear operator transforming the infinitesimal vectors $d\mathbf{X}$ into $d\mathbf{x}$ during the motion of the body defined by a differentiable global mapping (diffeomorphism) χ_κ of the manifold $\mathcal{B}_0$ on the three-dimensional Euclidean space ε_3. As a result, for each instant of time t a current configuration $\mathcal{B}_t$ of the body is reached. The deformation gradient **F** is a linear mapping defined on the so-called tangent space of the material manifold $\mathcal{B}_0$. The tangent space is represented by the material vectors essential for describing the body's deformation.

Choosing an arbitrary smooth curve $\mathcal{C}_0$ in the initial configuration $\mathcal{B}_0$, it is possible to investigate the $\mathcal{C}_t$ curve's current image. Let us write the equation of $\mathcal{C}_0$ in the parametric form

$$\mathbf{X} = \mathbf{X}(S), \tag{4.12}$$

where S is the parameter defining the path along the curve. Then the vector

$$\mathbf{T} = \frac{d\mathbf{X}}{dS} \tag{4.13}$$

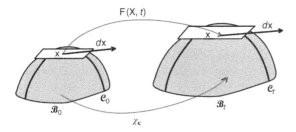

Figure 4.1 The own display of local configuration of a deformable continuous body, according to Wilmański (2008, Fig. 1, p. 5). *Source:* Copyright by Aleksandra Manecka – Padaż.

is the unit vector tangent to the curve. The infinitesimal vector

$$d\mathbf{X} = \mathbf{T}\, dS \tag{4.14}$$

is also tangent to curve $\mathcal{C}_0$. The infinitesimal vector $d\mathbf{x}$ is also tangential to the current image of the curve $\mathcal{C}_t$ due to the relations

$$d\mathbf{x} = \mathbf{t}\, dS, = \mathbf{FT} \tag{4.15}$$

Hence, the infinitesimal vector $d\mathbf{x}$ tangent to the curve $\mathcal{C}_t$, which deforms with the body, appears by a linear transformation of the infinitesimal vector $d\mathbf{X}$. Further details of the deformation of material vectors and deformation measures provide (Wilmański 1998), cf. Smith (1993). The polar decomposition of the deformation gradient $\mathbf{F}$ is suitable as a starting point for further discussion of the measures of finite deformation describing the change of the body's shape under the loading. Consider the neighbourhood of a chosen material point $\mathbf{X}$ and an infinitesimal material vector $d\mathbf{X}$ from the tangent space. The changes in the length of three such linearly independent vectors mimic the local changes in the geometry. In the current configuration $\mathcal{B}_t$, the actual length of the infinitesimal material vector reads:

$$ds^2 = d\mathbf{x} \cdot d\mathbf{x} = \left(\mathbf{F}d\mathbf{X}\right) \cdot \left(\mathbf{F}d\mathbf{X}\right) = \mathbf{C}\left(d\mathbf{X} \cdot d\mathbf{X}\right) \tag{4.16}$$

where

$$\mathbf{C} = \mathbf{F}^T\,\mathbf{F}, \mathbf{C}^T = \mathbf{C}, \tag{4.17}$$

is the right Cauchy–Green deformation tensor. The following polar decomposition theorem provides the interpretation of the above deformation measure (4.17)

$$\mathbf{F} = \mathbf{RU}, \mathbf{R}^{-1} = \mathbf{R}^T, \mathbf{U}^T = \mathbf{U}. \tag{4.18}$$

Due to (4.17) and (4.18), we have

$$\mathbf{C} = \left(\mathbf{RU}\right)^T \left(\mathbf{RU}\right) = \mathbf{U}^2, \tag{4.19}$$

where the symmetric tensor $\mathbf{U}$ is the right stretch tensor, and $\mathbf{R}$ is the tensor of rotation. The polar decomposition theorem also leads to the presentation of the deformation gradient with the inverse order:

$$\mathbf{F} = \mathbf{VR}, \mathbf{R}^{-1} = \mathbf{R}^T, \mathbf{V} = \mathbf{V}^T, \tag{4.20}$$

where $\mathbf{R}$ is the same orthogonal tensor appearing in the formula (4.18), and the symmetric tensor is the left stretch tensor. Equivalently, the relation for the left Cauchy–Green or Finger tensor $\mathbf{B}$ takes the form

$$\mathbf{B} = \left(\mathbf{VR}\right)\left(\mathbf{VR}\right)^{\mathrm{T}} = \mathbf{V}^2. \tag{4.21}$$

The time derivative of the relative deformation gradient taken for the actual configuration $\mathbf{F}_t(\mathbf{x}, \tau)$ leads to the relations

$$\mathbf{L} = \dot{\mathbf{F}}_t\left(\mathbf{x},\tau = t\right), \mathbf{D} = \frac{1}{2}\left(\mathbf{L} + \mathbf{L}^T\right), \mathbf{W} = \frac{1}{2}\left(\mathbf{L} - \mathbf{L}^T\right), \tag{4.22}$$

where $\mathbf{L}$ denotes the velocity gradient

$$\mathbf{L} = \mathbf{D} + \mathbf{W},$$

$\mathbf{D}$ is the symmetric stretching tensor (rate of deformation tensor) and $\mathbf{W}$ corresponds to the spin tensor.

References

Albers, B. and Wilmański, K. (2015). *Continuum Thermodynamics. Part II: Applications and Examples*. New Jersey, London, Singapore: World Scientific.

Bertram, A. (2014). On the History of Material Theory – A Critical Review. In: *Part II: Material Theories of Solid Continua and Solutions of Engineering Problems. The History of Theoretical, Material and Computational Mechanics – Mathematics Meets Mechanics and Engineering* (ed. E. Stein), 119–132. Berlin, Heidelberg: Springer-Verlag.

Coleman, B.D. and Noll, W. (1961). Foundations of linear viscoelasticity. *Rev. Mod. Phys.* 33: 239–243.

Frischmuth, K., Kosiński, W., and Perzyna, P. (1986). Remarks on mathematical theory of materials. *Arch. Mech.* 38: 59–69.

Hofstadter, D. (1979). *Gödel, Escher, Bach: An Eternal Golden Braid*. New York: Basic Books.

Ignatieff, Y.A. (1996). *The Mathematical World of Walter Noll. A Scientific Biography*. Berlin, Heidelberg: Springer-Verlag.

Man, C.-S. and Fosdick, R.L. (ed.) (2004). *The Rational Spirit in Modern Continuum Mechanics. Essays and Papers Dedicated to the Memory of Clifford Ambrose Truesdell III*. Dordrecht, Boston, MA, London: Kluwer Academic Publishers.

Meunier, M. (2012). *Industrial Applications of Molecular Simulations*. Boca Raton, FL, London, New York: CRC Press, Taylor & Francis.

Noll, W. (1972). A new mathematical theory of simple materials. *Arch. Ration. Mech. Anal.* 48: 1–50.

Noll, W. (1973). Lectures on the foundations of continuum mechanics and thermodynamics. *Arch. Ration. Mech. Anal.* 52: 62–92.

Olszak, W. and Ostrowska-Maciejewska, J. (1985). The plastic potential in the theory of anisotropic elastic-plastic solids. *Eng. Fract. Mech.* 4: 625–632.

Ostrowska-Maciejewska, J. (2020a). Rychlewski Jan. In: *Encyclopedia of Continuum Mechanics* (ed. H. Altenbach and A. Öchsner), 40–52. Berlin, Heidelberg: Springer-Verlag https://doi.org/10.1007/978-3-662-55771-6_344.

Ostrowska-Maciejewska, J. (2020b). *Phenomenological Mechanics of Continuous Media.* Warsaw: Published by Institute of Fundamental Technological Research, Polish Academy of Sciences (in Polish).

Perzyna, P. (1978). *Thermodynamics of Inelastic Solids.* Warszawa: PWN - Polish Scientific Publishers (in Polish).

Perzyna, P. (2010). The thermodynamical theory of elasto-viscoplasticity for a description of nanocrystalline metals. *Eng. Trans.* 58: 15–74. http://et.ippt.gov.pl.

Perzyna, P. (2012). Multiscale modelling of the influence of anisotropy effects on fracture phenomena in inelastic solids. *Eng. Trans.* 60: 225–284. http://et.ippt.gov.pl.

Perzyna, P. and Kosiński, W. (1973). A mathematical theory of materials. *Bull. Acad. Polon. Sci., Sér. Sci. Technol.* 21: 647–654.

Rychlewski, J. (1969). *Tensors and Tensor Functions.* Gdańsk: Institute of Fluid Flow Machinery, Polish Academy of Sciences. Bulletin No. 631. (in Polish).

Rychlewski, J. (1984). Elastic energy decomposition and limit criteria (originally in Russian). *Adv. Mech.* 7: 51–80; *Eng. Trans.* 59: 31–63, (2011) (in English).

Rychlewski, J. (1995a). Anisotropy and proper states of materials. In: *IUTAM Symposium on Anisotropy, Inhomogeneity and Nonliearity of Solids Mechanics* (ed. D.F. Parker and A.H. England), 19–24. Netherlands: Kluwer.

Rychlewski, J. (1995b). Unconventional approach to linear elasticity. *Arch. Mech.* 47: 149–171.

Rychlewski, J. (2000). A qualitative approach to Hooke's tensors. Part I. *Arch. Mech.* 52: 737–759.

Rychlewski, J. (2001). A qualitative approach to Hooke's tensors. Part II. *Arch. Mech.* 53: 45–63.

Smith, D.R. (1993). *An Introduction to Continuum Mechanics – After Truesdell and Noll.* Dordrecht: Kluwer Academic Publishers.

Truesdell, C. (1966). *Six Lectures on Modern Natural Philosophy.* Berlin, Heidelberg: Springer-Verlag.

Truesdell, C. (1969). *Six Lectures on Modern Natural Philosophy.* Warsaw: PWN – Polish Scientific Publishers (in Polish).

Truesdell, C. (1977). *A First Course in Rational Continuum Mechanics.* New York: Academic Press.

Truesdell, C. (1984). *Rational Thermodynamics,* 2e. New York, Berlin, Heidelberg, Tokyo: Springer-Verlag.

Truesdell, C. and Noll, W. (1992). *The Non-Linear Field Theories of Mechanics*, 2e. Berlin, Heidelberg: Springer-Verlag.

Truesdell, C. and Toupin. (1960). The classical field theories. In: *Encyclopedia of Physics* (ed. S. Flügge), 226–793. Berlin, Göttingen, Heidelberg: Springer-Verlag.

Wilmański, K. (1998). *Thermomechanics of Continua*. Berlin, Heidelberg: Springer-Verlag.

Wilmański, K. (2008). *Continuum Thermodynamics. Part I: Foundations*. New Jersey, London, Singapore: World Scientific.

5

Continuum Mechanics Description of Shear Banding

A detailed discussion about the physical picture of shear banding is provided in Chapter 3. It is visible that a more profound understanding of the
problem is still open and remains a tune of the future. Further research is
needed to develop a precise and complete derivation of material theory at
the atomic and microscopic levels of plastic glide vis-a-vis mesoscopic level
of micro-shear banding, leading to the macroscopic model of the viscoplastic flow. This question was the study's subject in the author's works, cf.
Pęcherski (1997, 1998a,b), Pęcherski and Korbel (2002), and Nowak and
Pęcherski (2002). Chapter 3 contains Hypothesis 3.1 about the extension of
the generally accepted concept of representative volume element (RVE).
The new original idea assumes the microscopic velocity field's singularity is
characterised with a jump of the velocity's tangential component.

5.1 System of Active Micro-shear Bands Idealised as the Surface of Strong Discontinuity

The physical constraint on any continuum mechanics approach to metal
plasticity, i.e. the dimension of crystalline material's smallest RVE, is the
subject of thorough discussion in Section 3.1. For the minimum RVE size,
one defines significant overall stress and strain measures during plastic flow.
The considerations are limited, typically, to relatively slow deformation processes with negligible body forces. Let us assume that within the reference
volume V_o of the macroscopic RVE (macro-element), the nominal stress
field $\mathbf{s}_m$ representing micro-stresses, and their rates $\dot{\mathbf{s}}_m$ are self-equilibrated,

$$Div\, \mathbf{s}_m = 0,\ Div\, \dot{\mathbf{s}}_m = 0 \text{ in } V_o, \tag{5.1}$$

and the following boundary conditions hold:

$$v_0 \mathbf{s}_m = \mathbf{t}_v,\ v_0 \dot{\mathbf{s}}_m = \dot{\mathbf{t}}_v,\ on\ \partial V_o, \tag{5.2}$$

where v_0 is the externally directed unit normal to the reference volume at a point on its boundary ∂V_o. The averaging procedure and micro-to-macro transition studied within the framework of finite deformation theory lead, in particular, to the following relations for the macroscopic measures of the deformation gradient $\mathbf{F}$ and its rate $\dot{\mathbf{F}}$, which are expressed, with the use of Gauss's theorem (divergence theorem), through surface data.

$$\mathbf{F} \equiv \{\mathbf{f}\} = \frac{1}{V_0} \int_{V_0} Grad\, \chi_m dV_o = \frac{1}{V_0} \int_{\partial V_o} x_m \otimes v_0 dA_0, \tag{5.3}$$

$$\dot{\mathbf{F}} \equiv \{\dot{\mathbf{f}}\} = \frac{1}{V_0} \int_{V_0} Grad\, \dot{\chi}_m dV_o = \frac{1}{V_0} \int_{\partial V_o} \dot{x}_m \otimes v_0 dA_0, \tag{5.4}$$

where the symbol χ_m denotes the microscopic field of motion of the material point X_m in the reference configuration of the RVE into its current position x_m,

$$x_m = \chi_m\left(X_m, t\right), \tag{5.5}$$

and the microscopic field of velocity $\mathbf{v}_m$ is determined in the current configuration,

$$\mathbf{v}_m = \mathbf{v}_m\left(x_m, t\right) = \mathbf{v}_m\left(\chi_m\left(X_m, t\right), t\right) \equiv \dot{\chi}_m\left(X_m, t\right). \tag{5.6}$$

Gauss's theorem applied was specified for any suitable vector field $\mathbf{w} = \mathbf{w}(X_m)$ defined on the closure $\overline{V}_o = V_o \cup \partial V_o$ and being of such a class that $\mathbf{w}$ is continuous on the closure $\overline{V}_o$ and continuously differentiable on V_o (cf., e.g. Smith 1993). Similarly, the following relations for the macroscopic measures of the appropriate smooth tensor field of nominal stress $\mathbf{S}$:

$$\mathbf{S} \equiv \{\mathbf{s}_m\} = \frac{1}{V_0} \int_{V_0} \mathbf{s}_m dV_o = \frac{1}{V_0} \int_{\partial V_o} X_m \otimes \mathbf{t}_v dA_0, \tag{5.7}$$

its rate $\dot{S}$ and $\dot{t}_v$

$$\dot{S} \equiv \left\{\dot{s}_m\right\} = \frac{1}{V_0} \int_{V_0} \dot{s}_m dV_o = \frac{1}{V_0} \int_{\partial V_o} \mathbf{X}_m \otimes \dot{t}_v dA_0, \tag{5.8}$$

and the Kirchhoff stress τ

$$\tau \equiv \left\{\tau_m\right\} = \left\{\mathbf{f}\mathbf{s}_m\right\} = \frac{1}{V_0} \int_{\partial V_o} \mathbf{x} \otimes \mathbf{t}_v dA_0, \tag{5.9}$$

can be obtained with the application of Gauss's theorem. The presented averaging procedure is valid under the assumption that plastic deformation's dominant mechanism corresponds to multiple crystallographic slips. In such a case, the theory describing kinematics and the constitutive structure of the finite elastic–plastic deformation of crystalline solids is well established. The transition between the microscopic and macroscopic levels is then well understood.

The preceding discussion of the shear banding physical nature and the results of microscopic observations *in situ* (Yang and Rey 1993, 1994) support Hypothesis 3.1 in Section 3.2 about the generalisation of the classical concept of RVE with the assumption of discontinuity surface of the microscopic velocity field. The adequate theory of propagating singular surfaces is available in the literature, e.g. Truesdell and Toupin (1960), Eringen and Suhubi (1974), and Kosiński (1986). The theory allows identifying the postulated discontinuity surface of the velocity field $\mathbf{v}_m$ in RVE as a singular surface $\Sigma(t)$ moving in the region V_o of the reference configuration of the body, where for each instant of 'time-like parameter' t the surface $\Sigma(t) \subset V_o$ has the dual counterpart $S(t) \subset V$ in the current configuration, Figure 5.1. There exist the jump discontinuity of derivatives of the function of motion χ_m, i.e. of the microscopic velocity field $\dot{\chi}_m \neq 0$ and the deformation gradient $[\![\mathbf{f}]\!] \neq 0$, which are assumed to be smooth in each point of $V_o \times I$, outside the discontinuity surface:

$$\left[\!\left[\dot{\chi}_m\right]\!\right] = \dot{\chi}_m^+ - \dot{\chi}_m^- \neq 0, \left[\!\left[\mathbf{f}\right]\!\right] = \mathbf{f}^+ - \mathbf{f}^- \neq 0, \tag{5.10}$$

where I is the range of '*time – like parameter t*', $I \subset R$. According to Truesdell and Toupin (1960) and Kosiński (1986), the considered surface of strong discontinuity of microscopic velocity field fulfils the properties of a vortex sheet with the jump discontinuity of the first derivatives of χ_m formally given by

$$\left[\!\left[\mathbf{v}_m\right]\!\right] = V_S s, \left[\!\left[\mathbf{v}_m\right]\!\right] = -\frac{V_S}{U} s \otimes nf, \text{ for } U \neq 0, \tag{5.11}$$

(a)   (b)

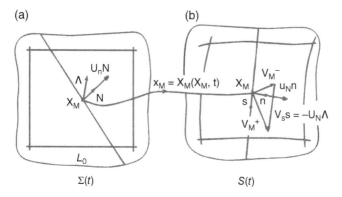

$\Sigma(t)$ $S(t)$

Figure 5.1 The dual representation of a strong discontinuity surface of tangential velocity jump traversing RVE: (a) in the reference configuration, (b) in the current configuration. *Source:* Copyright by Aleksandra Manecka – Padaż.

where s and n are the unit tangent and the unit normal vectors to the discontinuity surface $S(t)$, respectively, cf. Truesdell and Toupin (1960), p. 508, while according to Figure 5.1, $U = u_n - \mathbf{v}_m \cdot \mathbf{n}$ corresponds to the local speed of propagation of $S(t)$. Similarly, for the material counterpart of a singular surface $\Sigma(t)$, the compatibility relations read:

$$\llbracket \dot{\chi}_m \rrbracket = V_S s, \llbracket f \rrbracket = -\frac{V_S}{U_N} s \otimes N \text{ for } U_N \neq 0, \tag{5.12}$$

where N is the unit vector normal to the discontinuity surface $\Sigma(t)$ and U_N is the normal component of the surface velocity, cf. Eringen and Suhubi (1974), p. 96. Let us observe that the concept of vortex sheet remains of interest to fluid mechanics. It is a surface across which there is a discontinuity in fluid velocity, so slippage of one layer of fluid over another occurs. In contrast, the normal component of the flow velocity remains continuous. Therefore, using a vortex sheet idea typically on the shear surface or a shear layer in crystalline solids seems appropriate. Thus, the following observation transpires.

Observation 5.1

The micro-shear bands' initiation process accompanies local rotations of crystalline lattice in the adjacent material micro-volumes. The vorticity related to the vortex sheet concept is in accord with the intuitive understanding of the phenomenon typical for fluid mechanics. So the analogy also remains valid in the mechanics of solids, and the shear surface or a shear layer in crystalline solids holds.

The content of the above-mentioned statement in Observation 5.1 results from the discussion of Section 2.1.4. The physical nature of shear bands and the Observation 2.3 state that the local rotations of micro-material volumes and the misorientation of the crystalline lattice appear, cf. the discussion in Asaro (1979), Chang and Asaro (1980), and Lisiecki et al. (1982), where experimental data of local rotations of the crystalline lattice of fcc single crystals subjected to significant plastic straining are presented in detail.

5.1.1 On Finite Inelastic Deformations with High Lattice Misorientation

It is common to assume that dislocations traversing a volume element of the crystalline body produce a change of its shape but not its lattice orientation. It means that the material moves relative to its underlying crystalline lattice, while the lattice itself only undergoes elastic deformation. The situation is different because advanced deformations produce, locally, large lattice misorientation due to the fragmentation process. The concept of Somigliana dislocation finds application to describe theoretically the fragmented structure of strongly deformed crystalline solid near the onset of strain localisation in the course of shear banding. The importance of Somigliana dislocation for the theoretical modelling of different kinds of interfaces within the crystal was recognised by Eshelby (1961) and Asaro (1975) as well as Vergazov et al. (1977) and Romanov and Vladimirov (1983).

The construction of the Somigliana dislocation is as follows, cf. Eshelby (1961): cut the surface S and give the two faces of the cut an arbitrarily small relative displacement, removing material where there would be interpenetration, fill in the gaps, and weld the material together again. Let $B(x)$ be the relative displacement at the point x of the cut. Then we have a Somigliana dislocation over S specified by the displacement jump $B(x)$.

5.2 Macroscopic Averaging

The generalised form of Gauss's theorem (divergence theorem) for the gradient of the microscopic velocity field $\dot{\chi}_m$, sufficiently smooth in the RVE at each instant of the process is applied, i.e. in $V_o \times I$, besides the discussed singular surface, where the jump discontinuity appears $[\![\dot{\chi}_m]\!]$, leads to the following fundamental relation (Kosiński 1986), p. 68:

$$\frac{1}{V_0} \int_{V_0} Grad \, \dot{\chi}_m dV_o = \int_{\partial V_o - \Sigma(t)} \dot{\mathbf{x}}_m \otimes v_0 dA_0 - \int_{\Sigma(t)} [\![\dot{\chi}_m]\!] \otimes N dA_0. \tag{5.13}$$

Let us generalise the classical averaging procedure (4.19) of the microscopic velocity field $\dot{\chi}_m$ over the macroscopic RVE, which is traversed in the process of shear banding by a singular surface of order one. Then, the macroscopic measure of the deformation gradient $\mathbf{F}$ and its rate $\dot{\mathbf{F}}$ are defined through the surface data in the following way:

$$\mathbf{F} \equiv \frac{1}{V_0} \int_{\partial V_o - \Sigma(t)} \mathbf{x}_m \otimes v_0 dA_0, \tag{5.14}$$

and

$$\dot{\mathbf{F}} \equiv \frac{1}{V_0} \int_{\partial V_o - (t)} \dot{\chi}_m \otimes v_0 dA_0 = \frac{1}{V_0} \int_{v_0} Grad \, \dot{\chi}_m dV_o + \frac{1}{V_0} \int_{\Sigma(t)} [\![\dot{\chi}_m]\!] \otimes N dA_0. \tag{5.15}$$

Assuming the singular surface of order one with the velocity jump of the magnitude V_S and accounting for (5.12) transforms (5.15):

$$\dot{\mathbf{F}} = \frac{1}{V_0} \int_{V_0} Grad \, \dot{\chi}_m dV_o + \frac{1}{V_0} \int_{\Sigma(t)} V_S s \otimes N dA_0. \tag{5.16}$$

For the current configuration of RVE, at time t, chosen as the reference one, the rate of the deformation gradient $\dot{\mathbf{F}}$ becomes then the rate of the relative deformation gradient, $\dot{\mathbf{F}}_t(t)$, at time t, cf. (4.22), and the averaging formula (5.16) will take the following spatial form

$$\mathbf{L} = \frac{1}{V} \int_{\partial V - S(t)} \mathbf{v}_m \otimes v dA = \frac{1}{V} \int_V grad \, \mathbf{v}_m dV + \frac{1}{V} \int_{S(t)} V_S s \otimes n dA, \tag{5.17}$$

where $\mathbf{L}$ denotes the macroscopic measure of the velocity gradient

$$\mathbf{L} \equiv \dot{\mathbf{F}}_t(\mathbf{x}, t) = \dot{\mathbf{F}}(t) \mathbf{F}^{-1}(t), \tag{5.18}$$

averaged over the volume V of the current configuration of RVE swept out by the discontinuity surface $S(t)$. Similarly, application of the generalised form of the divergence theorem for the stress field $\mathbf{s}_m$ over the macro-element V_0 with the singular surface generates the formula for the average nominal stress

$$\mathbf{S} = \frac{1}{V_0} \int_{\partial V_0 - \Sigma(t)} \mathbf{X}_m \otimes t_v dA_0 = \frac{1}{V_0} \int_{V_0} \mathbf{s}_m dV_0 + \frac{1}{V_0} \int_{\Sigma(t)} \otimes [\![t_N]\!] dA_0, \tag{5.19}$$

where

$$\mathbf{t}_N = N\,\mathbf{s}_m \text{ and } \xi \in \Sigma\big(t\big).$$

The dynamical compatibility condition for the jump of the tractions $[\![\mathbf{t}_N]\!]$ across the singular surface in the reference configuration $\Sigma(t)$ reads, cf. Eringen and Suhubi (1974), p. 34:

$$N\ [\![\mathbf{s}_m]\!] = -\varrho U_N [\![\dot{\chi}_m]\!], \tag{5.20}$$

which specifies for a propagation discontinuity surface of tangential velocity jump in the spatial configuration, cf. Eringen and Suhubi (1974), p. 103:

$$\mathbf{n}\ [\![\tau_m]\!] = -\varrho U_N V_S \mathbf{s}. \tag{5.21}$$

Let us consider the processes in which the jump of inertial forces across the singular surface is negligible. It complies with the discontinuity surface movement as the progressing shear banding zone is near a quasi-static process. Then, $U = U_N = 0$ and the jump in the tractions $[\![\mathbf{t}_N]\!]$ vanishes to ensure the equilibrium condition. For $V_S = 0$ the known classical relation revealing no shear banding effects is retrieved.

$$\mathbf{L}_S = \frac{1}{V}\int_V grad\ \mathbf{v}_m dV \tag{5.22}$$

where the index 'S' underlines that the velocity gradient $\mathbf{L}_S$ describes the inelastic deformations mediated by the crystallographic slip mechanism.

The averaging formula (5.17) enables us to account for the contribution of micro-shear banding in the macroscopic measure of velocity gradient produced at finite strain. It is visible that the total velocity gradient $\mathbf{L}$ decomposes in the following way

$$\mathbf{L} = \mathbf{L}_S + \mathbf{L}_{SB}, \ L_{SB} = \frac{1}{V}\int_{S(t)} V_S \mathbf{s} \otimes \mathbf{n} dA. \tag{5.23}$$

Assuming that the singular surface $S(t)$ forms a plane traversing volume V, with the unit vectors $\mathbf{s}$ and $\mathbf{n}$ held constant, (5.23) results in

$$\mathbf{L}_{SB} = \dot{\gamma}_{SB}\mathbf{s} \otimes \mathbf{n}, \tag{5.24}$$

where the macroscopic shear strain rate $\dot{\gamma}_{SB}$ determines the microscopic variables as an average over the RVE,

$$\frac{1}{V} \int_{S(t)} H_{ms} \dot{\gamma}_{MS} dA = \frac{1}{V} \int_{S(t)} H_{ms} B_{ms} \rho_{ms} v_{ms} dA. \tag{5.25}$$

Assuming for simplicity that the structural parameter B_{ms} and the speed v_{ms} are constant over the surface S(t), we have

$$\dot{\gamma}_{SB} = B_{ms} \rho_{SB} v_{ms}, \tag{5.26}$$

where

$$\rho_{SB} = \frac{1}{V} \int_{S(t)} H_{ms} \rho_{ms} dA. \tag{5.27}$$

The symbol ρ_{SB} denotes the macroscopic volume density of micro-shear bands that operate within the sequence of clusters sweeping the RVE. The density ρ_{SB} may change with the 'time-like parameter' t for the magnitudes H_{ms} and ρ_{ms} are, in general, varied for different clusters. If we assume the average rate of change $\dot{\rho}_{SB}$ with the amount of the density ρ_{SB} of shear bands operating in the period $\Delta\tau$, has the meaning as an infinitesimal increment of 'time-like parameter' in the macroscopic description, the following equivalent form takes place

$$\dot{\gamma}_{SB} = B_{ms} L_{ms} \dot{\rho}_{SB}. \tag{5.28}$$

The derived relations are valid for a single system of micro-shear bands. This situation can be generalised for the case of a double shear banding system

$$\mathbf{L} = \mathbf{L}_S + \sum_{i=1}^{2} \dot{\gamma}_{SB}^{(i)} \mathbf{s}^{(i)} \otimes \mathbf{n}^{(i)}, i = 1, 2, \tag{5.29}$$

where $\dot{\gamma}_{SB}^{(i)}$ is the macroscopic shear strain rate and $\mathbf{s}^{(i)}, \mathbf{n}^{(i)}$ are the respective unit vectors of the i th shear banding system. The relation (5.29) provides the following macroscopic measures of the rate of inelastic deformations and spin produced by active micro-shear bands.

$$\mathbf{D}_{SB} = \frac{1}{2}\left(\mathbf{L}_{SB} + \mathbf{L}_{SB}^{T}\right), \mathbf{W}_{SB} = \frac{1}{2}\left(\mathbf{L}_{SB} - \mathbf{L}_{SB}^{T}\right) \tag{5.30}$$

It is worth observing that (5.29) is valid under the simplifying assumption that the active shear bands in both systems operate simultaneously in time, corresponding to a sufficiently small increment of 'time-like parameter' in the macroscopic description. Otherwise, more realistically, one should account for the cyclic sequence of events. Observe that Korbel et al. (1999, 2003) provided a new mathematical description of sequential activation of single shear banding generating finite plastic deformation. The procedure of limit transition enables obtaining the analytical solution for simultaneous double shearing, which is free from constitutive assumptions.

Let us recall the observation revealing two types of shear banding mechanisms generating the inelastic deformation in materials.

- The first type of shear banding corresponds to *the rapid formation of the multiscale shear banding systems*. The micro-shear band clusters propagate and produce the discontinuity of microscopic velocity field v_m. A new concept of the RVE with a strong singularity appears, and the **instantaneous shear banding contribution function f_{SB}** originates. In such a case, the simplifying assumption shows that the active shear bands in both systems operate simultaneously in time, i.e. in a sufficiently small increment of 'time-like parameter' in the macroscopic description.
- The second type of shear banding relates, on the other hand, to *a progressive kind of shear band* based on the accumulation of the particular contribution of micro-shear bands forming clusters gradually. Finally, they accumulate in the localisation zone spreading across the macroscopic volume of considered material. Such a deformation mechanism appears in amorphous solids as glassy metals or polymers and metallic solids of nanocrystalline structures. It seems that there are local shear transformation zones (STZs), cf. Argon (1979, 1999), and Scudino et al. (2011). The STZ clusters of randomly close-packed atoms that cooperatively reorganise under applied shear stress represent the fundamental units of plasticity in metallic glasses. Their percolation initiates and contributes to the *cumulative kind of shear banding*. Then, the **volumetric contribution function f_{SB}^v** of shear banding appears and enters the competitive game of viscoplastic flow. In such a case, the more realistic assumption transpires that the active shear bands in both systems are somewhat distant from the immediate ones and are operating sequentially. Thus, the cyclic activation of single shear banding generates finite inelastic deformation.

An example of numerical simulations of the tensile tests of a hypothetical rectangular sample is considered. As a result, the deformation gradient takes the multiplicative form, composed of subsequent slip contributions, and it appears to be non-commutative concerning the order of shearing events. It transpires that the final shape of the hypothetical sample depends on the order in which the various plastic shears are applied. The alternating shear strain increments $\Delta\gamma_1 = \Delta\gamma_2$ produce the rotation angle φ of two initially orthogonal edges of the considered sample. The illustration is displayed in Figure 5.2, where each point of the diagram represents the final configuration of the specimen, obtained after the total shear strain $\sum(\Delta\gamma_1 + \Delta\gamma_2) = 1$. The

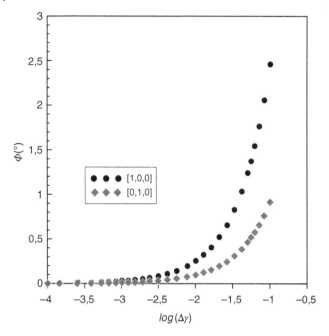

Figure 5.2 The effect of shear strain increments, taken in the logarithmic scale, during the sequential operation of two symmetrically disposed shear banding systems upon the rotation of the two initially orthogonal edges lying in the planes of Miller indices [1,0,0] and [0,1,0] of a hypothetical rectangular sample, cf. Korbel et al. (2003), p. 50, Fig. 1. *Source:* Copyright by Ryszard Pęcherski.

shear strain γ_1 resulting in the single shear process produces the deformation gradient $\mathbf{F}_1 = \mathbf{1} + \gamma_1 \mathbf{b}_1 \otimes \mathbf{n}_1$. In the case of the sequence of two slips, the initial material element $\mathbf{l}_0$ transforms as $\mathbf{l}_1 = \mathbf{F}_1 \mathbf{l}_0$ and, subsequently, takes the position of $\mathbf{l}_2 = \mathbf{F}_2 \cdot \mathbf{l}_1$, where $\mathbf{F}_2 = \mathbf{1} + \gamma_2 \mathbf{b}_2 \otimes \mathbf{n}_2$. Similarly, this has resulted in the sequence of n shears (Korbel et al. 1999, 2003). The model developed herein applies particularly for the numerical analysis of the rotation of the tensile axis within the finite plastic deformation kinematics due to the cyclic sequence of slips in face cubic centred (FCC) single crystal. The trajectory of the trace of the tensile axis on a stereographic projection is analysed. The results are compared with similar calculations of Shalaby and Havner (1978) made for double slip, cf. also Havner (1992). In the limit transition, for diminishing shear strain increments, both predictions converge. Furthermore, the numerical simulation shows that the tensile axis repeatedly passes the symmetry line $\left[001\right]-\left[1\,\bar{1}1\right]$, which finds confirmation in the experimental observations of overshoot phenomena but not with the double-slip theory, cf. Korbel et al. (2003) p. 354, Fig. 3.

References

Argon, A.S. (1979). Plastic deformation in metallic glasses. *Acta Metall.* 27: 47–58.

Argon, A.S. (1999). Rate processes in plastic deformation of crystalline and noncrystalline solids. In: *Mechanics and Materials: Fundamentals and Linkages* (ed. M.A. Meyers, R.W. Armstrong and H. Kirchner), 175–230. New York: Wiley.

Asaro, R.J. (1975). Somigliana dislocations and internal stresses with application to second phase hardening. *Int. J. Eng. Sci.* 13: 271–286.

Asaro, R.J. (1979). Geometrical effects in the inhomogeneous deformation of ductile single crystals. *Acta Metall.* 27: 445–453.

Chang, Y.W. and Asaro, R.J. (1980). Lattice rotations and localized shearing in single crystals. *Arch. Mech.* 32: 369–388.

Eringen, A.C. and Suhubi, E.S. (1974). *Elastodynamics: Finite Motions*, vol. 1. New York: Academic Press.

Eshelby, J.D. (1961). Elastic inclusions and inhomogeneities. In: *Progress in Solid Mechanics*, vol. II (ed. I.N. Sneddon and R. Hill), 89–140. Amsterdam: North Holland.

Havner, K.S. (1992). *Finite Plastic Deformation of Crystalline Solids*. Cambridge: Cambridge University Press.

Korbel, K., Pęcherski, R.B., and Korbel, A. (1999). Analysis of the tensile axis rotation of a FCC single crystal within the framework of the kinematics of plastic deformation by sequential slip (in Polish). *Rudy i Metale Nieżelazne* 44 (11): 542–545.

Korbel, K., Pęcherski, R.B., and Korbel, A. (2003). Analysis of finite plastic deformation due to the sequence of slips. *Voprosy Materialovedeniâ - Problems of Materials Science* 33 (2): 349–356.

Kosinski, W. (1986). *Field Singularities and Wave Analysis in Continuum Mechanics*. Chichester: PWN – Polish Scientific Publishers, Warszawa and Elis Horwood Limited Publishers.

Lisiecki, L.L., Nelson, D.Q., and Asaro, R.J. (1982). Lattice rotations, necking and localized deformation in FCC single crystals. *Scr. Metall.* 16: 441–448.

Nowak, Z. and Pęcherski, R.B. (2002). Plastic strain in metals by shear banding. II. Numerical identification and verification of plastic flow law. *Arch. Mech.* 54: 621–634.

Pęcherski, R.B. (1997). Macroscopic measure of the rate of deformation produced by micro-shear banding. *Arch. Mech.* 49: 385–401.

Pęcherski, R.B. (1998a). Macroscopic effects of micro-shear banding in plasticity of metals. *Acta Mech.* 131: 203–224.

Pęcherski, R.B. (1998b). Macromechanical description of micro—shear banding. In: *Proceedings of the McNU '97 Symposium on Damage Mechanics in Engineering Materials, Studies in Applied Mechanics*, June 28 — July 2, 1997, Evanston, vol. 46 (ed. G.Z. Voyiadjis, J.W. Ju and J.-L. Chaboche), 203–222. New York: Elsevier.

Pęcherski, R.B. and Korbel, K. (2002). Plastic strain in metals by shear banding. I. Constitutive description for simulation of metal shaping operations. *Arch. Mech.* 54: 603–620.

Romanov, A.E. and Vladimirov. (1983). Disclinations in solids. *Phys. Status Solidi A* 78: 11–34.

Scudino, S., Jerliu, B., Pauly, S. et al. (2011). Ductile bulk metallic glasses produced through designed heterogeneities. *Scr. Mater.* 65: 815–818.

Shalaby, A.H. and Havner, K.S. (1978). A general kinematical analysis of double slip. *J. Mech. Phys. Solids* 26: 19–92.

Smith, D.R. (1993). *An Introduction to Continuum Mechanics – After Truesdell and Noll*. Dordrecht: Kluwer Academic Publishers.

Truesdell, C. and Toupin, R.A. (1960). The classical field theories. In: *Encyclopedia of Physics* (ed. S. Flügge), 226–793. Berlin, Göttingen, Heidelberg: Springer-Verlag.

Vergazov, A.N., Likhachev, V.A., and Rybin, V.V. (1977). Characteristic elements of the dislocation structure in deformed polycrystalline molybdenum. *Phys. Met. Metall.* 42: 126–133 (Russian).

Yang, S. and Rey, C. (1993). Analysis of deformation by shear banding: a two-dimensional post-bifurcation model. In: *Large Plastic Deformations*, 277–278. Routledge.

Yang, S. and Rey, C. (1994). Shear band postbifurcation in oriented copper single crystals. *Acta Metall.* 42: 2763–2774.

6

Deformation of a Body Due to Shear Banding – Theoretical Foundations

6.1 Basic Concepts and Relations of Finite Inelastic Deformation of Crystalline Solids

Consider a polycrystalline aggregate as a continuum body. An infinitesimal neighbourhood of a material point $\mathbf{X}$ of the body is discussed in Chapter 3 with the use of the concept of representative volume element (RVE). It is sufficient for an accurate continuum mechanics description of gross elastic–inelastic behaviour, i.e. viscoplastic one for the rate-dependent yield strength or a plastic one in rate-independent deformation processes. The dominant orientation of the crystalline lattice in the RVE is represented by the triad of director vectors. We can choose an arbitrary triad of orthogonal unit vectors serving as a reference. It was the renowned scholar Jean Mandel (1907–1982) who, following the seminal work of Cosserat brothers (Cosserat and Cosserat 1909) on polar continua, introduced a 'trièdre directeur' (Mandel 1971, 1973). Some applications of the Mandel concept recall, for instance, Raniecki and Nguyen (1984), Cleja-Tigoiu and Soós (1990), and Pęcherski (1983, 1985), where more detailed discussion and further references are available. In the plasticity of crystalline solids, one usually assumes that dislocations traversing a volume element produce a shape change, but they do not variate its lattice orientation. It plays the crucial role of being the local intermediate, relaxed configuration idea of finite plastic deformation of polycrystalline aggregate. The originality of the idea conceived by Mandel (1971, 1973) lies in the isoclinic property of the local intermediate configuration. It means that the chosen director triad always keeps the same orientation concerning the laboratory reference frame's fixed axes. Consequently, the decomposition of the deformation gradient $\mathbf{F}$

Viscoplastic Flow in Solids Produced by Shear Banding, First Edition. Ryszard B. Pęcherski.
© 2022 John Wiley & Sons Ltd. Published 2022 by John Wiley & Sons Ltd.

$$\mathbf{F} = \mathbf{EP}, \tag{6.1}$$

becomes unique, while $\mathbf{E}$ denotes the elastic transformation from the intermediate isoclinic configuration to the current one, and $\mathbf{P}$ is the inelastic transformation, accordingly viscoplastic or plastic one, from the reference configuration to the isoclinic one. The similar decomposition of the deformation gradient $\mathbf{F}$ also belongs to Lee (1969) and Lee and Liu (1967) but without explicit definition of a structure or director vectors and isoclinic configuration. Nevertheless, the following fundamental kinematical relations hold:

$$\mathbf{L} = \dot{\mathbf{E}}\mathbf{E}^{-1} + \mathbf{E}\dot{\mathbf{P}}\mathbf{P}^{-1}\mathbf{E}^{-1} \tag{6.2}$$

$$\mathbf{D} = \mathbf{D}^e + \mathbf{D}^p, \tag{6.3}$$

where

$$\mathbf{D}^e = \left\{\dot{\mathbf{E}}\mathbf{E}^{-1}\right\}_s \tag{6.4}$$

is the elastic part, while

$$\mathbf{D}^p = \left\{\mathbf{E}\dot{\mathbf{P}}\mathbf{P}^{-1}\mathbf{E}^{-1}\right\}_s \tag{6.5}$$

denotes the inelastic part of the rate of deformation. Similarly,

$$\mathbf{W}^e = \left\{\dot{\mathbf{E}}\mathbf{E}^{-1}\right\}_a, \tag{6.6}$$

is the elastic part of the spin, and

$$\mathbf{W}^p = \left\{\mathbf{E}\dot{\mathbf{P}}\mathbf{P}^{-1}\mathbf{E}^{-1}\right\}_a \tag{6.7}$$

denotes the inelastic part of the spin. The symbols $\{\mathbf{t}\}_s$ and $\{\mathbf{t}\}_a$ correspond to the symmetric and skew-symmetric parts of the second-order tensor. In the literature, the historical origin and priority of the ideas above are under discussion. Besides the Mandel concept of the intermediate isoclinic configuration, the similar thought of Eckart (1948) is raised by Kleiber and Raniecki (1985) and Perzyna (1978), as well as Clayton (2011). The novel concepts of other prominent scholar Ekkehart Kröner (1919–2000), cf. e.g. Kröner (1959, 1961) are discussed in Perzyna (1978), Chakrabarty (2000), Nemat-Nasser (2004), and Gurtin et al. (2009), as well as Khan and Huang (1995) and Clayton (2011). The comprehensive and fundamental study on nonlinear continuum theory of dislocations and internal stresses within the framework of the general Riemann–Cartan differential geometry and generalised Cosserat continuum led Kröner to the following idea of three configurations of an elastic–plastic body element (Kröner 1959) p. 285: 'der Ideal - oder Anfangszustand, der natürliche oder Zwischenzustand, der deformirte oder Endzustand' (*ideal or initial state, natural or intermediate state, deformed or final state*) and to the application of '*triedre mobile*' of the Cosserat brothers, Kröner (1959), p. 292. At the end of the story, it is worth emphasising the origins thread of the multiplicative decomposition of the deformation gradient presented by Clayton (2011) and Souhayl and Yavari (2015). Besides the contributions mentioned above, the first attempts of Kondo (1952, 1955) as well as Bilby et al. (1955, 1957) and Sedov (1965) are worth recalling. The first formal introduction of the multiplicative decomposition of the deformation gradient in finite plasticity appeared in the above works realised independently that a body in plastic deformation could be relaxing in a stress-free intermediate configuration.

6.2 Continuum Model of Finite Inelastic Deformations with Permanent Lattice Misorientation

Let us assume that in the current configuration of the continuous body, the vector fields $\mathbf{e}_i$, $i = 1, 2, 3$ suffer discontinuity on certain surfaces. The concept of Somigliana dislocation can model these surfaces. Furthermore, three groups of translational dislocations within the body are assumed, cf. Bilby (1960) and Kröner (1959): those that leave it or annihilate in it, which remain continuously distributed which remain on the surface of the discontinuity. The latter corresponds to the known interpretation of the Somigliana dislocation as infinitesimal translational dislocations continuously distributed on the surface S. Movement of translational dislocations produces irreversible deformation. In contrast, their accumulation on the boundaries contributes to the local rotations of the microvolumes. It creates permanent crystal lattice misorientation. The question then arises about incorporating this new effect into the continuum theory of finite inelastic deformations of a crystalline body. The extension of the notion of the Burgers circuit for disclinations, discussed by Lardner (1973, 1974), appears helpful to solve this problem. See the presentation of this issue provided in Pęcherski (1983, 1985), also mentioned by Clayton (2011) and Clayton et al. (2006). Figure 6.1 displays the intuitive visualisation of the lattice misorientation orthogonal tensor $\mathbf{R}_m(C)$ and the original Lardner's construction of a smooth loop C of unloading neighbourhoods displaying the idea of the Burgers circuit for Somigliana dislocation.

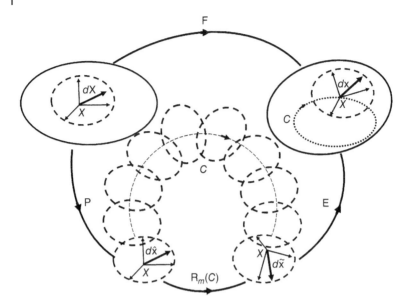

Figure 6.1 Configurations of a single crystal during the finite elastic and inelastic deformations with lattice misorientation. Visualisation of a smooth circuit C of unloading neighbourhoods defining the misorientation tensor $\mathbf{R}_m(C)$. *Source:* Copyright by Ryszard Pęcherski.

Consider the reference configuration of a crystalline body and the current configuration in the inelastic deformation process under applied loads. Imagine then in the actual configuration a particular piecewise smooth circuit C. Let us suppose that the local unloading configuration exists for each neighbourhood of the point from the circuit C and that such a neighbourhood can be chosen in a continuous way, cf. Figure 6.1. If two such regions overlap, then we require that two local unloading configurations fit together without any free rotation. We can construct the unloading configuration sequence at points along the circuit starting with the local unloading configuration orientation, the same as the reference one. We can do this by selecting a sequence of overlapping neighbourhoods along the circuit and then choosing configurations continuously for adjacent regions. The question arises whether the unloading configuration we end up with is the same as we started. If the configuration is not the same, the change due to going around a circuit C can only consist of a closure failure and relative rigid-body rotation, for the local unloading configurations are, by definition, in a natural state. In such a case, the discontinuity surface $S(t)$ corresponds to Somigliana dislocation. The relative rigid body rotation pertains to the lattice misorientation at the point x and is represented by the orthogonal tensor $\mathbf{R}_m(C)$. Following the above-mentioned procedure with the circuit C, the deformation gradient $\mathbf{F}$ reveals the new decomposition (Pęcherski 1983, 1985):

$$\mathbf{F} = \mathbf{ER}_m\left(C\right)\mathbf{P}, \tag{6.8}$$

while $\mathbf{E}$ denotes the elastic transformation from the intermediate isoclinic configuration to the misoriented one, and $\mathbf{P}$ is the inelastic transformation from the reference configuration to the isoclinic one.

If making the circuit C produces closure failure only, without rigid body rotation, the Somigliana dislocation degenerates into a dislocation of the translational type, and decomposition (6.8) takes the form of (6.1). Let us consider the deformation of the neighbourhood of a material point $\mathbf{X}$ of the body with the position vector $d\,\mathbf{X}$. In the translation, dislocations glide process with the Burgers vector $d\,\mathbf{b}_g$ pass through the vector $d\,\mathbf{X}$, the latter transforms into its intermediate configuration $d\,\mathbf{x}$, cf. Figure 6.2.

$$d\mathbf{x}_g = d\mathbf{X} + d\mathbf{b}_g. \tag{6.9}$$

or with an account of the transformation

$$d\mathbf{x}_g = \mathbf{P}d\mathbf{X} \tag{6.10}$$

$$d\mathbf{b}_g = \left(\mathbf{P} - \mathbf{1}\right)d\mathbf{X}. \tag{6.11}$$

Then the local relative rigid body rotation transforms the position vector $d\mathbf{X}$ into its configuration $d\mathbf{x}_r$ as follows

$$d\mathbf{x}_r = \mathbf{R}_m\left(C\right)\mathbf{P}d\mathbf{X} \tag{6.12}$$

and

$$d\mathbf{b}_r = d\mathbf{x}_r - d\mathbf{x}_g = \left(\mathbf{R}_m\left(C\right) - \mathbf{1}\right)\mathbf{P}d\mathbf{X}, \tag{6.13}$$

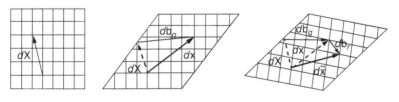

Figure 6.2 Deformation of the material point **X** neighbourhood with the position vector $d\,\mathbf{X}$ in the inelastic deformation and the lattice misorientation. *Source:* Copyright by Ryszard Pęcherski.

where $d\,\mathbf{b}_r$ is the increment of the Burgers vector produced by the local relative lattice rotation determined by the misorientation tensor $\mathbf{R}_m(C)$ at point $\mathbf{x}$. Thus, the total relative lattice rotation of the Burgers vector $d\,\mathbf{B}_t$ is given by

$$d\mathbf{B}_t = d\mathbf{b}_g + d\mathbf{b}_r. \tag{6.14}$$

Due to the last relations, we have

$$d\mathbf{B}_t = \left(\mathbf{R}_m\left(C\right)\mathbf{P} - \mathbf{1}\right)d\mathbf{X} \tag{6.15}$$

and the rate of the Burgers vector increment $d\,\mathbf{B}_t$ reads

$$d\dot{\mathbf{B}}_t = \left(\dot{\mathbf{R}}_m\left(C\right)\mathbf{P} + \mathbf{R}_m\left(C\right)\dot{\mathbf{P}}\right)d\mathbf{X} \tag{6.16}$$

or owing to transformation (6.12), the following relation corresponding to the intermediate configuration takes the form

$$d\dot{\mathbf{B}}_t = \left(\dot{\mathbf{R}}_m\left(C\right)\mathbf{R}_m^{-1}\left(C\right) + \mathbf{R}_m\left(C\right)\dot{\mathbf{P}}\mathbf{P}^{-1}\mathbf{R}_m^{-1}\left(C\right)\right)d\mathbf{x}_r \tag{6.17}$$

and connects of the total Burgers vector increment $d\dot{\mathbf{B}}_t$ with the continuum measures of inelastic distortion and lattice misorientation rates. Let us observe a detailed discussion on the physical basis of Somigliana dislocation and the Burgers circuit C enclosing a surface element ΔS by Pęcherski (1985), assuming that in the limit transition, the value of ΔS is diminishing, and the dependency of the misorientation tensor $\mathbf{R}_m(C)$ on C can be neglected. Then, due to the expression for the velocity gradient (5.18), the decomposition of the gradient of deformation (6.8) leads to the accounting for the lattice misorientation velocity gradient:

$$\mathbf{L} = \dot{\mathbf{E}}\mathbf{E}^{-1} + \mathbf{E}\dot{\mathbf{R}}_m\mathbf{R}_m^{-1}\mathbf{E}^{-1} + \mathbf{E}\mathbf{R}_m\mathbf{P}\dot{\mathbf{P}}^{-1}\mathbf{R}_m^{-1}\mathbf{E}^{-1}. \tag{6.18}$$

Taking the symmetric and antisymmetric parts of (6.18), we obtain

$$\mathbf{D} = \left(\dot{\mathbf{E}}\mathbf{E}^{-1}\right)_s + \left(\mathbf{E}\dot{\mathbf{R}}_m\mathbf{R}_m^{-1}\mathbf{E}^{-1}\right)_s + \left(\mathbf{E}\mathbf{R}_m\mathbf{P}\dot{\mathbf{P}}^{-1}\mathbf{R}_m^{-1}\mathbf{E}^{-1}\right)_s \tag{6.19}$$

for the total rate of deformation and

$$\mathbf{W} = \left(\dot{\mathbf{E}}\mathbf{E}^{-1}\right)_a + \left(\mathbf{E}\dot{\mathbf{R}}_m\mathbf{R}_m^{-1}\mathbf{E}^{-1}\right)_a + \left(\mathbf{E}\mathbf{R}_m\mathbf{P}\dot{\mathbf{P}}^{-1}\mathbf{R}_m^{-1}\mathbf{E}^{-1}\right)_a \tag{6.20}$$

for the total spin.The inelastic parts, plastic or viscoplastic ones, of the rates of deformation and spin read

$$\mathbf{D}^p = \left(\mathbf{E}\dot{\mathbf{R}}_m\mathbf{R}_m^{-1}\mathbf{E}^{-1}\right)_s + \left(\mathbf{E}\mathbf{R}_m\mathbf{P}\dot{\mathbf{P}}^{-1}\mathbf{R}_m^{-1}\mathbf{E}^{-1}\right)_s, \tag{6.21}$$

$$\mathbf{W}^p = \left(\mathbf{E}\dot{\mathbf{R}}_m \mathbf{R}_m^{-1} \mathbf{E}^{-1}\right)_a + \left(\mathbf{E}\mathbf{R}_m \dot{\mathbf{P}} \mathbf{P}^{-1} \mathbf{R}_m^{-1} \mathbf{E}^{-1}\right)_a. \tag{6.22}$$

To the inelastic rate of deformation the parts produced due to the mechanisms of crystallographic slip $\mathbf{D}_S$ and shear banding $\mathbf{D}_{SB}$ contribute:

$$\mathbf{D}^p = \mathbf{D}_S + \mathbf{D}_{SB} \tag{6.23}$$

where we have as a result of crystallographic slip

$$\mathbf{D}_S = \left(\left(\left(\mathbf{E}\dot{\mathbf{R}}_m \mathbf{R}_m^{-1} \mathbf{E}^{-1}\right)_s + \left(\mathbf{E}\mathbf{R}_m \dot{\mathbf{P}} \mathbf{P}^{-1} \mathbf{R}_m^{-1} \mathbf{E}^{-1}\right)_s\right)\right)_S \tag{6.24}$$

and the macroscopic shear banding produces the following inelastic rate of deformation

$$\mathbf{D}_{SB} = \left(\left(\left(\mathbf{E}\dot{\mathbf{R}}_m \mathbf{R}_m^{-1} \mathbf{E}^{-1}\right)_s + \left(\mathbf{E}\mathbf{R}_m \dot{\mathbf{P}} \mathbf{P}^{-1} \mathbf{R}_m^{-1} \mathbf{E}^{-1}\right)_s\right)\right)_{SB}. \tag{6.25}$$

Similarly, the inelastic part of the spin reads:

$$\mathbf{W}^p = \mathbf{W}_S + \mathbf{W}_{SB}, \tag{6.26}$$

where due to the crystallographic slip, we have:

$$\mathbf{W}_S = \left(\left(\left(\mathbf{E}\dot{\mathbf{R}}_m \mathbf{R}_m^{-1} \mathbf{E}^{-1}\right)_a + \left(\mathbf{E}\mathbf{R}_m \dot{\mathbf{P}} \mathbf{P}^{-1} \mathbf{R}_m^{-1} \mathbf{E}^{-1}\right)_a\right)\right)_S, \tag{6.27}$$

and the macroscopic shear banding produces the following inelastic spin

$$\mathbf{W}_{SB} = \left(\left(\left(\mathbf{E}\dot{\mathbf{R}}_m \mathbf{R}_m^{-1} \mathbf{E}^{-1}\right)_a + \left(\mathbf{E}\mathbf{R}_m \dot{\mathbf{P}} \mathbf{P}^{-1} \mathbf{R}_m^{-1} \mathbf{E}^{-1}\right)_s\right)\right)_{SB}. \tag{6.28}$$

The additive decomposition (6.23) of the rate of inelastic deformation leads to the scalar shear banding contribution function:

$$f_{SB} = \frac{\|\mathbf{D}_{SB}\|}{\|\mathbf{D}^p\|}, \tag{6.29}$$

where the norm of a symmetric second-order tensor $\mathbf{A}$ reads

$$\|A\| = \sqrt{A\,A} = \sqrt{A_{ij}A_{ij}}. \tag{6.30}$$

We shall call the defined ratio in (6.29) the *instantaneous shear banding contribution function* f_{SB}.

As discussed in Pęcherski and Korbel (2002), among many possible realisations of shear banding, one can single out the group of processes characterised with the exact contribution of two symmetric shear banding systems. In the case of proportional loading paths, the total inelastic shear strain rate can be expressed as the sum of shear strain rate mediated by crystallographic slip $\dot{\gamma}_S$ and shear banding $\dot{\gamma}_{SB}$:

$$\dot{\gamma} = \dot{\gamma}_S + \dot{\gamma}_{SB.} \tag{6.31}$$

Due to (6.29), we have equivalently:

$$f_{SB} = \frac{\dot{\gamma}_{SB}}{\dot{\gamma}}, \tag{6.32}$$

which results in:

$$\dot{\gamma}\left(1 - f_{SB}\right) = \dot{\gamma}_S, \tag{6.33}$$

where

$$f_{SB} \in \left[0,1\right). \tag{6.34}$$

6.3 Basic Concepts and Relations of Constitutive Description – Elastic Range

For the sake of brevity and simplicity, the sole isothermal processes became the subject of interest. The development of the thermodynamic theory of inelastic materials can refer to Nemat-Nasser (2004) as well as Gurtin et al. (2009) and Asaro and Lubarda (2006). Assume the mechanical state variables $(\pi, \mathbf{A})$ corresponding to the isoclinic configuration, where π is the second Piola–Kirchhoff stress tensor related to the Cauchy stress:

$$\pi = \left(det\,\mathbf{E}\right)\mathbf{E}^{-1}\sigma\mathbf{E}^{T}, \tag{6.35}$$

and $\mathbf{A}$ represents the set of internal variables. The elastic Green strain

$$\Delta^{e} = \frac{1}{2}\left(\mathbf{E}^{T}\mathbf{E} - 1\right) \tag{6.36}$$

can be calculated from the free enthalpy function H per unit mass, which may take the form

$$H = H\left(\pi,\mathbf{A}\right) = H_1\left(\pi\right) + H_2\left(\mathbf{A}\right), \tag{6.37}$$

$$\Delta^e = -\rho_k \frac{\partial H}{\partial \pi},$$ (6.38)

where $H = \Phi - \Delta^e : (\pi/\rho_k)$ provides the relation with the free energy function per unit mass, Φ, depending in case of isothermal processes in the following particular form on the state variables

$$\left(\Delta^e, \mathbf{A}\right), \phi\left(\Delta^e, \mathbf{A}\right) = \phi_1\left(\Delta^e\right) + \phi_2\left(\mathbf{A}\right).$$ (6.39)

The studies of Raniecki and Nguyen (1984) and Nguyen (1992) show it is typical that the distortional elastic strain of metallic solids remains small under arbitrary loading conditions, while it can undergo significant reversible dilatational changes in shape under very high pressure. Applying the earlier concept of Rice (1975), who noted that the tensor of elastic moduli in Eulerian description is expressed in terms of derivatives of the free energy as simple as in the case of infinitesimal strains, provided the logarithmic elastic strain $\varepsilon = ln\,\mathbf{V}^e$ is adopted as a state variable and that the values of the ratios of principal elastic stretches $\mathbf{U}^e$, from the polar decomposition $\mathbf{E} = \mathbf{V}^e\mathbf{R}^e = \mathbf{R}^e\mathbf{U}^e$, belong to the interval [5/6, 7/6]. In this way, the authors extended the earlier results of Willis (1969), obtained under the stronger assumption that the yield stress in simple shear is much less than the elastic shear modulus. Let us confine the further analysis to small distortional and dilatational elastic strains. Then, the following approximate relation appears within the accuracy of $O\left(|e|^2\right)$, where e is the deviatoric part of ε, cf. Raniecki and Nguyen (1984) and Nguyen (1992) for more detailed discussion:

$$\mathbf{D}^e = \overset{\circ}{\varepsilon} = \dot{\varepsilon} + \varepsilon\mathbf{W}^e - \mathbf{W}^e\varepsilon, \mathbf{W}^e = \dot{\mathbf{R}}^e\dot{\mathbf{R}}^{e\mathbf{T}}$$ (6.40)

with the elasticity equation

$$\overset{\circ}{\tau} = \mathbf{C} : \mathbf{D}^e, = \rho_k \frac{\partial^2 \phi}{\partial\varepsilon\partial\varepsilon}, \overset{\circ}{\tau} = \dot{\tau} + \tau\mathbf{W}^e - \mathbf{W}^e\tau,$$ (6.41)

where $\mathbf{C}$ is the fourth-order tensor of elastic moduli and τ is the Kirchhoff stress tensor.

6.4 The Yield Limit Versus Shear Banding – The 'extremal surface'

Let us refer to the pivotal papers of Hill (1967, 1979) to discuss the meaning of 'yield' within the context of micro-shear banding. The precise connection of the nominal yield points with intrinsic material properties was discussed earlier by Hill (1967, 1979), who proposed the idea of an 'extremal surface'. Imagine that after a given prestrain of the RVE of a polycrystalline aggregate, further glide hardening on its constituent grains' active slip systems is suspended. In general, due to constraint hardening, the incremental inelastic flow under constant overall load is still precluded. This approach can apply to the microscopic fields of stress and strain within the RVE. According to Hill (1967), such fields correspond to intrinsic eigenstates. The micro-shear bands result from the particular configuration of internal micro-stresses that accumulate at grain boundaries until the grain hardening on the active slip systems is suspended and then abruptly released. Then, under constant overall load, the field of inelastic deformation rate as a self-induced deformation mode is produced. In this way, the analogy with the intrinsic eigenstates by Hill (1967, 1979) is confirmed. Such a view is in accord with experimental observations discussed previously in Chapter 2. (Hill 1967), which emphasises that the 'extremal surface' is not a single surface but is instead an assemblage of yield points as the envelope about physically distinct states of RVE. None of these states can reach any other via purely elastic paths in the stress space. The following observation correlates the 'extremal surface' properties with the mechanism of the onset of new micro-shear band systems.

Observation 6.1

The 'extremal surface' forms the generic micro-shear banding surface that we call the instantaneous envelope of plastic states produced by shear banding. The properties of the 'extremal surface' conform with the multiscale mechanism of micro-shear band systems activation. The yield conditions sweeping particular positions on the 'extremal surface' pertain to micro-shear bands' subsequent spatial patterns. The transition from one state to the other one is not possible via a purely elastic path. Instead, specific accumulated plastic strain is necessary to produce the new set of micro-shear bands characterised, in general, by another geometric pattern.

In Chapter 8, the pertinent discussion of the plastic flow model with the external surface accounting for the onset of shear banding visualising the above-mentioned 'extremal surface' and the internal surface relating to the elastic range and back stress anisotropy with nonlinear evolution of kinematic hardening is provided.

References

Asaro, R.J. and Lubarda, V.A. (2006). *Mechanics of Solids and Materials*. Cambridge, New York: Cambridge University Press.

Bilby, B.A. (1960). Continuous distribution of dislocations. In: *Progress in Solid Mechanics*, vol. 1 (ed. I.N. Sneddon and R. Hill), 332. Amsterdam: North-Holland.

Bilby, B., Bullough, R., and Smith, E. (1955). Continuous distributions of dislocations: a new application of the methods of non-riemanian geometry. *Proc. R. Soc. London, Ser. A* 231: 263–273.

Bilby, B.A., Lardner, L.R.T., and Stroh, A.N. (1957). Continuous distributions of dislocations and the theory of plasticity. *Actes du IXe congrés international de mécanique appliquée. Bruxelles, 1956* 8: 35–44.

Chakrabarty, J. (2000). *Applied Plasticity*. New York: Springer-Verlag.

Clayton, J.D. (2011). *Nonlinear Mechanics of Crystals*. Dordrecht, Heidelberg, London, New York: SpringerScience+Business Media B.V.

Clayton, J.D., McDowell, D.L., and Baumann, D.J. (2006). Modeling dislocations and disclinations with finite micropolar elastoplasticity. *Int. J. Plast.* 22: 210–256.

Cleja-Tigoiu, S. and Soós, E. (1990). Elastoviscoplastic models with relaxed configurations and internal state variables. *Appl. Mech. Rev.* 43: 131–151.

Cosserat, E. and Cosserat, F. (1909). *Théorie des corps déformables. Hermann.*

Eckart, G. (1948). Theory of elasticity and inelasticity. *Phys. Rev.* 73: 373–380.

Gurtin, M.E., Fried, E., and Anand, L. (2009). *The Mechanics and Thermodynamics of Continua*. Cambridge, New York (Hardback): Cambridge University Press.

Hill, R. (1967). The essential structure of constitutive laws for metal composites and polycrystals. *J. Mech. Phys. Solids* 15: 779–795.

Hill, R. (1979). Theoretical plasticity of textured aggregates. *Math. Proc. Cambridge Philos. Soc.* 85: 179–191.

Khan, A.S. and Huang, S. (1995). *Continuum Theory of Plasticity*. New York, Chichester, Brisbane, Toronto, Singapore: Wiley.

Kleiber, M. and Raniecki, B. (1985). Elastic-plastic materials at finite strains. In: *Plasticity Today: Modelling, Methods and Applications* (ed. A. Sawczuk and G. Bianchi), 3–46. London and New York: Elsevier Applied Science Publishers.

Kondo, K. (1952). On the geometrical and physical foundations of the theory of yielding. *Proc. Jpn. Nat. Cong. Appl. Mech.* 2: 41–47.

Kondo, K. (1955). Non-Riemannian geometry of imperfect crystals from a macroscopic viewpoint. In: *RAAG Memoirs of the Unifying Study of Basic Problems in Engineering and Physical Science by Means of Geometry*, vol. I (ed. K. Kondo). Tokyo: Gakuyusty Bunken Fukin-Kay.

Kröner, E. (1959). Allgemeine Kontinuumstheorie der Versetzungen und Eigenspannungen. *Arch. Rat. Mech. Anal.* 4: 18–334.

Kröner, E. (1961). Zur Plastischen Verfornung des Vielkristalls. *Acta Metall.* 9: 155–161.

Lardner, R.W. (1973). Foundations of the theory of disclinations. *Arch. Mech.* 25: 911–922.

Lardner, R.W. (1974). *Mathematical Theory of Dislocations and Fracture.* Toronto: University of Toronto Press.

Lee, E.H. (1969). Elastic-plastic deformation at finite strain. *J. Appl. Mech.* 36: 1–6.

Lee, E. and Liu, D. (1967). Finite-strain elastic-plastic theory with application to plane-wave analysis. *J. Appl. Phys.* 38: 19–27.

Mandel, J. (1971). *Plasticité et visciplasticité*, CISM Lecture Notes No. 97. Udine: Springer. Wien.

Mandel, J. (1973). Equations constitutives et directeurs dans les milieux plastiques et viscoplastiques. *Int. J. Solids Struct.* 9: 725–740.

Nemat-Nasser, S. (2004). *Plasticity. A Treatise on Finite Deformation of Heterogeneous Inelastic Materials.* Cambridge: Cambridge University Press.

Nguyen, H.V. (1992). Constitutive equations for finite deformations of elastic-plastic metallic solids with induced anisotropy. *Arch. Mech.* 44: 585–594.

Pęcherski, R.B. (1983). Relation of microscopic observations to constitutive modelling for advanced deformations and fracture initiation of viscoplastic materials. *Arch. Mech.* 35: 257–277.

Pęcherski, R.B. (1985). Discussion of sufficient condition for plastic flow localisation. *Eng. Fract. Mech.* 21: 767–779.

Pęcherski, R.B. and Korbel, K. (2002). Plastic strain in metals by shear banding. I. Constitutive description for simulation of metal shaping operations. *Arch. Mech.* 54: 603–620.

Perzyna, P. (1978). *Termodynamika Materiałów Niesprężystych (Thermodynamics of Inelastic Materials).* Warszawa: PWN.

Raniecki, B. and Nguyen, H.V. (1984). Isotropic elastic-plastic solids at finite strain and arbitrary pressure. *Arch. Mech.* 36: 687–704.

Rice, J. (1975). Continuum mechanics and thermodynamics of plasticity in relation to microscale deformation mechanics. In: *Constitutive equations of plasticity* (ed. A.S. Argon), 23–79. Cambridge, MA: MIT Press.

Sedov, L.I. (1965). *Introduction to the Mechanics of a Continuous Medium (first published in Russian in 1962).* Reading, MA: Addison-Wesley.

Souhayl, S. and Yavari, A. (2015). On the origins of the idea of the multiplicative decomposition of the deformation gradient. *Math. Mech. Solids* 22: 771–772.

Willis, J.R. (1969). Some constitutive equations applicable to problems of large dynamic plastic deformation. *J. Mech. Phys. Solids* 17: 359–369.

7

The Failure Criteria Concerning the Onset of Shear Banding

7.1 The Yield Condition for Modern Materials – the State of the Art

The advances in the recent studies of rational mechanics of materials are manifested, among others, in the relationship to strive for a better understanding of the deformation mechanisms and physical failure processes in newly designed and manufactured materials. The analysis of many industrial and laboratory applications should account for some unique mechanical and functional properties of investigated solids. One should distinguish the tailored directional properties by the multiscale control of the structure on different observation levels. Also, the asymmetry of elastic range produced by multilevel processes of shear banding (Pęcherski 1997, 1998a,b; Pęcherski et al. 2011, 2014) is possible to observe. The asymmetry of elastic range transpires as the difference of tensile and compression strength, the so-called strength differential effect (SDE). Modern mechanics of solids attempt to provide such a mathematical description, enabling us to account for these features. The purpose of this chapter is to formulate a well-founded yield condition based on the elasticity limit criterion for solids revealing the asymmetry of elastic range.

It appears physically justified to assume that the considered range of elastic limit depends on the concept of elastic energy density stored in the strained solids. The advantage of energetic measures of material effort lies in the multiscale character of the energy concept. It means that a *measure of material effort* has a well-defined physical interpretation in each scale under consideration.

The analysis of a deformable crystalline body across scales leads to the above definition. According to fundamentals of the solid-state physics presented, e.g. by Phillips (2001), Gilman (2003), or Finnis (2003), the density

Viscoplastic Flow in Solids Produced by Shear Banding, First Edition. Ryszard B. Pęcherski.
© 2022 John Wiley & Sons Ltd. Published 2022 by John Wiley & Sons Ltd.

Definition 7.1

The material effort characterises the deformed body with the strength of chemical bonds in the natural state.

---

and spatial distribution of valence electrons control the elastic stiffness and the strength of a solid body. The values of these quantities are evaluated in terms of the total energy of the considered atomic system. According to the detailed discussion in Pęcherski et al. (2014), the energy determined on the microscopic scale relates to the deformed body's elastic energy density considered on the macroscopic level as a continuum. (Phillips 2001), p. 245, shows that:

$$\frac{1}{2}C_{ijkl}\varepsilon_{ij}\varepsilon_{kl} = \frac{1}{\Omega}\left[E_{tot}\left(\left\{R_i^{def}\right\}\right) - E_{tot}\left(\left\{R_i^{undef}\right\}\right)\right], \tag{7.1}$$

where C_{ijkl} denote the components of the stiffness tensor, ε_{ij} is the measure of infinitesimal elastic strains, while Ω is the considered volume element in which the total energy of the deformed state $E_{tot}\left(\left\{R_i^{def}\right\}\right)$ evaluated on the microscopic level and subtracting off the energy of the undeformed (natural) state $E_{tot}\left(\left\{R_i^{undef}\right\}\right)$ equilibrates the elastic energy density on the continuum level. The symbols $\left\{R_i^{def}\right\}$ and $\left\{R_i^{undef}\right\}$ correspond to the sets of ions in the volume Ω of the considered solid body in the deformed and undeformed states, respectively.

This definition is based on the earlier studies of the relation of microscopic observations to the modelling of viscoplastic flow and failure of solids by Pęcherski (1983), as well as on the applications of molecular dynamics for the analysis of the strength of chemical bonds determining, as an example, the strength of metal and metal–oxide interfaces studied in Nalepka and Pęcherski (2009) and Nalepka et al. (2015). The historical account of using the elastic energy density as the material effort measure is provided in work (Pęcherski et al. 2014).

Experiments proved a definite influence of pressure on the yield stress, cf., e.g. Richmond and Spitzig (1980), Spitzig et al. (1976), and Wilson (2002). It is worthwhile to stress that in particular (Wilson 2002) underlined the importance of the Bridgman pivotal investigations (Bridgman 1947, 1952) leading to the paradigm change, cf. the related discussion in Vadillo et al. (2011). Due to the spread of popular, sometimes contradictory, opinions on the effects of pressure in classical metal plasticity, the citation of Wilson (2002), p. 63, seems to be well founded: 'Bridgman's earlier work

(Bridgman 1947), showed no systematic effect of pressure on strain harden-ing. However, his later work was characterized by more precise measure-ments which established a definite effect of hydrostatic pressure on the strain hardening curves of mild steel'. And further concerning the earlier observations of Bridgman (1947), p. 63: '. . . Bridgman found that the mate-rial volume in the gage section did not change for very large plastic strain changes. Therefore, a metal was assumed to have incompressible plastic strains. These two experimental observations about metal – no influence of hydrostatic pressure on yielding and incompressibility for plastic strain changes – are two of the basic tenets of classical metal plasticity'.

The distinct difference of the yield stress values as a function of pressure shows that the asymmetry of the elastic range appears. The strength differ-ential effect's origin is deeply rooted in the quantum mechanical basis of electronic material structure. It is visible in the calculations of the UBER function (universal binding energy relation), discussed, for instance, in the example of Cu by Nalepka (2013) and presented in Pęcherski et al. (2014). UBER provides the physical explanation why most of the considered mate-rials reveal the elastic range's asymmetry, i.e. the limit of strength in com-pression is higher than the limit of strength in tension. In this way, the UBER function justifies the assumptions of the material effort hypothesis (Burzyński 1928, 2009). Rose et al. (1984) observed using a vast set of *ab initio* data obtained for different metals that the total energy of a metal crys-tal per atom changes in the process of uniform volume expansion according to the following general formula:

$$E\left(r\right) = -E_{coh}\left[1 + \eta\left(\left(\frac{V}{V_0}\right)^{1/3} - 1\right)\right]e^{-\eta\left(\left(\frac{V}{V_0}\right)^{1/3} - 1\right)}, \tag{7.2}$$

where

$$\eta = \sqrt{9B_0 V_0 / E_{coh}} \tag{7.3}$$

denotes the scaling parameter, while V_0, B_0, and E_{coh} are the equilibrium volume per atom, bulk modulus, and cohesion energy (in positive value). Additionally, V constitutes the actual crystal volume per atom. For cubic metals, the ratio $\left(V / V_0\right)^{1/3}$ reduces to the ratio of lattice constants a/a_0. For example, the UBER curve for copper reconstructed by a semi-empirical potential (Nalepka 2013) is presented (Figure 7.1). The function matches the *ab initio* data (Mishin et al. 2001) very well.

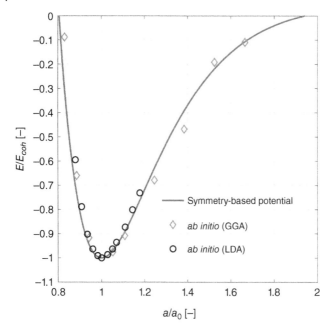

Figure 7.1 UBER function for copper reconstructed by symmetry-based Voter potential (Nalepka 2013) compared with the *ab initio* data (Mishin et al. 2001). *Source:* Copyright by Kinga Nalepka.

The isotropic body complete criterion formulation was provided originally by Burzyński (1928, 2009) and Vadillo et al. (2011) as regards more recent results.

7.2 The Yield Condition for the Isotropic Materials Revealing the Strength Differential Effect

The discussion above concludes that the concept of energy may find application in a universal measure of material effort. A particular part of the atom system's energy relates to the change of chemical bonds' strength due to the shift in energy between interatomic interactions. Similarly, on the macroscopic level, some precisely defined contributions of the density of elastic energy accumulated in the strained body account for the measure of material effort. The energy measure of material effort proposed by Burzyński (1928, 2009) reflects the varying contributions of the volumetric and distortional parts of elastic energy density correctly in the deformation process.

This new idea of an energy-based approach also confirms that of other authors. For instance, Freudenthal (1950), p. 20, states: 'Correlation of behaviour of the different levels is possible only in terms of concept which on all levels has the same meaning in both Newtonian and statistical mechanics, the same dimension, and the same tensorial rank. This concept is energy'. On the other hand, the need for a valid account for the interplay between both parts of elastic energy density, deviatoric and volumetric ones, is underlined by Christensen (2004, 2006), cf. also for more detailed discussion (Pęcherski 2008).

Burzyński presented the hypothesis of variable limit energy of volume change and distortion:

Hypothesis 7.1

'The measure of local material effort in elastic and plastic ranges is the sum of density of quasi-energy of distortion and a certain part – dependent on the state of stress and individual properties of a body – of the density of the pseudo-energy of volume change' – Burzyński (2009), p. 141.

Adding the prefixes 'quasi' or 'pseudo' emphasises that the analytic expressions used in continuation do not mean – for a particular group of bodies or relatively in specific experimental fields – elastic energy in the sense of the linear theory of elasticity.

The mathematical formulation of the hypothesis of variable limit energy of volume change and distortion reads (Burzyński 1928, 2009):

$$\Phi_f + \eta \Phi_v = K, \quad \eta = \omega + \frac{\delta}{3p}, \quad p = \frac{\sigma_1 + \sigma_2 + \sigma_3}{3}, \tag{7.4}$$

where Φ_f denotes the density of elastic energy of distortion

$$\Phi_f = \frac{1}{12G}\left[\left(\sigma_1 - \sigma_2\right)^2 + \left(\sigma_1 - \sigma_3\right)^2 + \left(\sigma_2 - \sigma_3\right)^2 \right], \tag{7.5}$$

and Φ_v is the elastic energy density of volume change

$$\Phi_v = \frac{1-2v}{6E}\left(\sigma_1 + \sigma_2 + \sigma_3\right)^2 = \frac{1-2v}{12G(1+v)}\left(\sigma_1 + \sigma_2 + \sigma_3\right)^2 \tag{7.6}$$

expressed in terms of principal stresses $\sigma_1 \geq \sigma_2 \geq \sigma_3$ and contributes to the total elastic energy for isotropic solid

$$\Phi = \Phi_f + \Phi_v, \tag{7.7}$$

while ω and δ are material constants, which are to be specified, and $p = \dfrac{\sigma_1 + \sigma_2 + \sigma_3}{3}$ denotes mean stress expressed in principal stress coordinates. Furthermore, ν represents Poisson's ratio, and G is the Kirchhoff shear modulus. By introducing the function η, Burzyński took into account the experimentally based information that the increase of mean stress p results in the diminishing contribution of the elastic energy density of volume change in the measure of material effort. Accounting for the above relations leads to the following final form of the Burzyński failure hypothesis:

$$\frac{1}{3}\sigma_f^2 + 3\frac{1-2\nu}{1+\nu}\omega p^2 + \frac{1-2\nu}{1+\nu}\delta p = 4GK, \tag{7.8}$$

where $\sigma_f^2 = 12G\Phi_f$. The main idea of Burzyński's derivation lies in the conversion of the unknown material parameters (ω, δ, K) into the triplet (k_t, k_c, k_s) relatively easy to obtain in the laboratory tests: elastic (plastic) limit obtained in uniaxial tension – k_t, in uniaxial compression – k_c, and in torsion – k_s, respectively. A more detailed discussion of this derivation and resulting failure criteria is given in Burzyński (1928, 2009) and in the recent papers (Frąś and Pęcherski 2010; Frąś et al. 2010). The replacement of the material parameters (ω, δ, K) with the material constants (k_t, k_c, k_s) in (7.8) leads to the following relation:

$$\frac{k_c k_t}{3k_s^2}\sigma_e^2 + \left(9 - \frac{3k_c k_t}{k_s^2}\right)p^2 + 3\left(k_c - k_t\right)p - k_c k_t = 0, \quad p = \frac{\sigma_1 + \sigma_2 + \sigma_3}{3}, \tag{7.9}$$

where

$$\sigma_e = \frac{1}{\sqrt{2}}\sqrt{\left[\left(\sigma_1 - \sigma_2\right)^2 + \left(\sigma_1 - \sigma_3\right)^2 + \left(\sigma_2 - \sigma_3\right)^2\right]} \tag{7.10}$$

denotes equivalent stress used in the theory of plasticity. According to the detailed study presented in Burzyński (1928, 2009), the Eq. (7.9) in the space of principal stresses $\sigma_1 \geq \sigma_2 \geq \sigma_3$ represents particular cases of quadric surfaces depending on the interplay between three material constants (k_t, k_c, k_s). Burzyński considered the following basic modes of stress:

I)   Uniaxial tension: $\sigma_1 = k_t, \sigma_2 = 0, \sigma_3 = 0$
II) Uniaxial compression: $\sigma_1 = 0, \sigma_2 = 0, \sigma_3 = -k_c$
III) Torsion (shear): $\sigma_1 = k_s, \sigma_2 = 0, \sigma_3 = -k_s$
IV) Biaxial uniform tension: $\sigma_1 = k_{tt}, \sigma_2 = k_{tt}, \sigma_3 = 0$

V) Biaxial uniform compression: $\sigma_1 = 0$, $\sigma_2 = -k_{cc}$, $\sigma_3 = -k_{cc}$
VI) Triaxial uniform tension: $\sigma_1 = k_{ttt}$, $\sigma_2 = k_{ttt}$, $\sigma_3 = k_{ttt}$
VII) Triaxial uniform compression: $\sigma_1 = -k_{ccc}$, $\sigma_2 = -k_{ccc}$, $\sigma_3 = -k_{ccc}$.

The first example of a quadric, expressed by the general relation (7.9), corresponds, in the three-dimensional system of principal stress coordinates, to an ellipsoid of revolution of the symmetry axis $\sigma_1 = \sigma_2 = \sigma_3$, for $k_s > \sqrt{\dfrac{k_c k_t}{3}}$.

The second example corresponds, in the three-dimensional system of principal stress coordinates, to paraboloid of revolution of the symmetry axis $\sigma_1 = \sigma_2 = \sigma_3$, for $k_s = \sqrt{\dfrac{k_c k_t}{3}}$. Then, the general formula (7.9) leads to the following equation:

$$\sigma_e^2 + 3\left(k_c - k_t\right)p - k_c k_t = 0. \tag{7.11}$$

The third case corresponds, in the three-dimensional system of principal stress coordinates, to hyperboloid of revolution for the symmetry axis $\sigma_1 = \sigma_2 = \sigma_3$, which stems from the general formula (7.9) for $\dfrac{2}{\sqrt{3}}\dfrac{k_c k_t}{k_c + k_t} < k_s < \sqrt{\dfrac{k_c k_t}{3}}$. Generally, the surface has two separate sheets, but only one sheet of the hyperboloid comes into play from a practical perspective. In the plane coordinate system (σ_f, p), the criterion is depicted by a hyperbola. There is a particular case: $k_s = \dfrac{2}{\sqrt{3}}\dfrac{k_c k_t}{k_c + k_t}$, then the hyperboloid degenerates to a rotationally symmetric conical surface. One sheet of the surface accounts for the description of the limit criterion obtained from (7.9) in the following way:

$$\sigma_e + 3\frac{k_c - k_t}{k_c + k_t}p - 2\frac{k_c k_t}{k_c + k_c} = 0. \tag{7.12}$$

The above-mentioned cases of limit surface – ellipsoid, paraboloid, and conical one – are depicted in the plane coordinate system (σ_e, p) (Figure 7.2).

The yield condition of the form given in (7.12) and depicted in Figure 7.2 has been widely known in the literature since 1952 as Drucker–Prager yield criterion cf. Drucker and Prager (1952). It is worthwhile to note, however, that some authors have recognised the earlier contribution of Burzyński (1928, 2009) – for instance: Sendeckyj (1972), Pisarenko and Lebedev (1973), as well as Jirásek and Bažant (2002), Nardin et al. (2003), Yu (2004), and Yu et al. (2006). The above-discussed yield or strength criteria obtained rigorously from the energy-based material effort hypothesis proposed originally

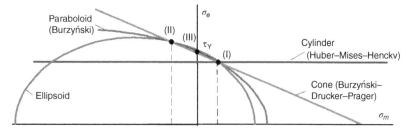

Figure 7.2 Illustration of different cases of Burzyński quadrics depicted in the plane coordinate system (σ_e, p), where $\sigma_m \equiv p$, cf. Frąś (2013). *Source:* Copyright by Teresa Frąś.

by Burzyński in 1928 were later re-discovered several times for different empirical parameters independently by many researchers, cf. Pęcherski (2008) for the confrontation with the works of other authors.

7.3 Examples and Visualisations of the Particular Burzyński Failure Criteria

Some examples of the visualisations of the discussed above limit criteria are displayed. The experimental investigations mainly carrying in the plane state of stress provide the required data. The yield surface in the three-dimensional space of principal stresses is reconstructed (Wolfram Mathematica, 6.0 Copyright 1988–2007). The mentioned experimental tests are related usually with the uniaxial modes of stresses I and II, the mode of pure shear III as well as with the biaxial ones IV and V for the plane states of stress inscribed in specific hypothetical yield limit of material with the thought strength differential ratio $\kappa = k_c / k_t = 1.3$, Figure 7.3.

The two particular specifications of the Burzyński Hypothesis 7.1 related with the ellipsoidal and paraboloidal surfaces find an illustration for a chosen material.

7.3.1 Ellipsoidal Failure Surface

An example of the material revealing the before-discussed inequality $k_c < k_t$ is, e.g. the magnesium alloy AZ31, for which, according to the data published in Yoshikawa et al. (2008): the yield strength in tension $k_t = 200$ MPa, and the yield strength in compression $k_c = 120$ MPa, while the yield strength in shear $k_s = 120$ MPa, under the quasi-static test conditions with the offset strain

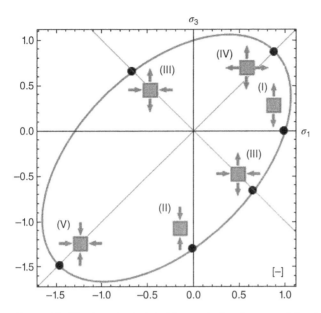

Figure 7.3 The hypothetical yield curve in the plane $\sigma_2 = 0$ for the case of strength differential ratio $\kappa = 1.3$ with schematically indicated points corresponding to the results of particular laboratory tests. *Source:* Copyright by Teresa Frąś.

0.002. The yield surface appears as an ellipsoid in the principal stress coordinates, Figure 7.4a, and the ellipse in the plane $\sigma_2 = 0$, Figure 7.4b. While Figure 7.4c depicts the half-ellipse in the plane coordinate system (σ_e, p) and Figure 7.4d shows the comparison of the symmetric Huber–Mises criterion displaying in the blue dotted line with the trace of the ellipsoid yield surface in the plane $\sigma_2 = 0$. The limit curves in Figures 7.4b–d contain the above-mentioned experimental measurements' points in black, cf. Frąś (2013).

7.3.2 Paraboloid Failure Surfaces

The historical experimental investigations results for various materials were collected by Theocaris (1995). The failure surfaces' visualisations resulted, cf. Figure 7.5a–d. The first case reveals a very distinct strength differential effect measured by the ratio $\kappa = 3$ corresponding to grey cast-iron investigations (Grassi and Cornet 1949; Coffin 1950), cf. Frąś (2013).

Another example is related to the yield limit's first investigations in the complex states of stress (Lode 1926; Taylor and Quinney 1931). Theocaris (1995) collected the results and expressed them in non-dimensional form. The failure paraboloid surfaces visualisations are depicted in Figure 7.6a–d.

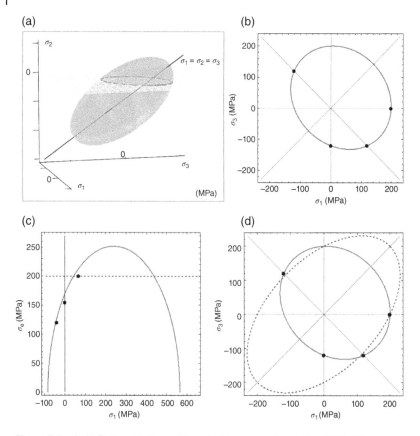

Figure 7.4 (a, b) Failure limit condition for the magnesium alloy AZ31: (a) ellipsoid yield surface in the principal stress coordinates and (b) the ellipse in the plane $\sigma_2 = 0$. Source: Copyright by Teresa Frąś. (c, d) Failure limit condition for the magnesium alloy AZ31: (c) the half-ellipse in the plane coordinate system (σ_e, p) and (d) the Huber–Mises criterion in the blue dotted line confronting the trace of the ellipsoid yield surface in the plane $\sigma_2 = 0$. *Source:* Copyright by Teresa Frąś.

The visible deviation of experimental results from the prediction according to the Huber–Mises yield condition might account for the strength differential effect measured by the ratio $\kappa = 1.3$. This discrepancy could explain that the mentioned deviation can also result from slight initial anisotropy produced by manufacturing material elements, e.g. drawing rods and machining samples, cf. Frąś (2013).

The following example is related to the innovative nanocrystalline (nc) metals investigated in numerical simulations by Lund and Schuh (2003) and Schuh and Lund (2003). The authors performed molecular simulations of multiaxial deformation of metallic glass using an energy minimisation technique. The results reveal the asymmetry between the magnitudes of the yield strength in tension and compression, with the strength difference ratio

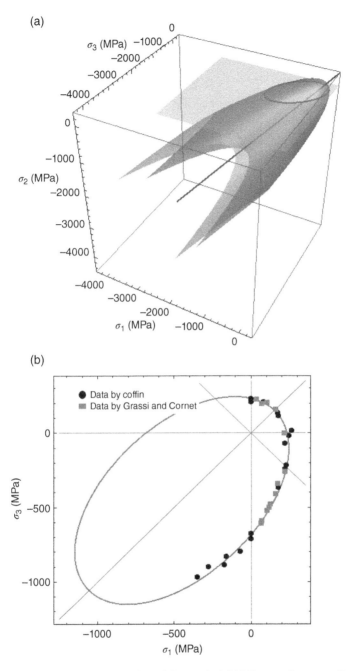

Figure 7.5 (a, b) Visualisation of the paraboloid failure surface according to the data of Grassi and Cornet (1949) and Coffin (1950) for grey cast iron: (a) in the three-dimensional space of principal stresses, (b) in the plane $\sigma_2 = 0$. Source: Copyright by Teresa Frąś. (c, d) Visualisation of the paraboloid failure surface according to the data of Grassi and Cornet (1949) and Coffin (1950) for grey cast iron: (c) in the coordinate system (σ_e, p), (d) comparison with the Huber–Mises criterion presented in the plane $\sigma_2 = 0$. Source: Copyright by Teresa Frąś.

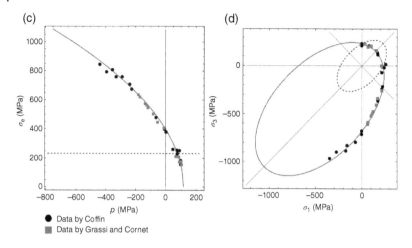

Figure 7.5 (Continued)

$\kappa = 1.24$. At ambient temperature, metallic glasses deform in the process of shear banding, where plastic strains localise into strips (micro-bands) of nanometer thickness. The mechanism of shear banding on the nanometer-scale produces unique mechanical properties at the macroscopic scale. For example, the low tensile elongation values recorded for amorphous metals lead to rapid failure and a single shear band. In constrained loading modes like compression, plastic yielding appears serrated, and a sequence of single shear bands produces the strain. These properties became the subject of investigations in various glassy alloys with different compositions, and the observation appears to be general to this class of materials. One crucial consequence of shear localisation in amorphous metals is that the macroscopic yield criterion may depend on the maximum shear stress and the hydrostatic pressure or the normal stress acting on the shear plane. The discussion and proposal of a model accounting for shear-banding mechanism in ufg and nc metals are provided (Nowak et al. 2007; Frąś et al. 2011). The results displayed in Figure 7.7a,b show that the Burzyński paraboloidal yield condition correlates satisfactorily with the discussed Lund and Schuh (2003) and Schuh and Lund (2003) data, cf. Frąś (2013).

7.4 Remarks on the Extension Including Anisotropic Materials

In his pivotal paper, Mises (1928) had presented a thorough study of limit criteria for anisotropic solids. Olszak and Ostrowska-Maciejewska (1985) and Rychlewski (1984) formulated the energy-based foundations of the

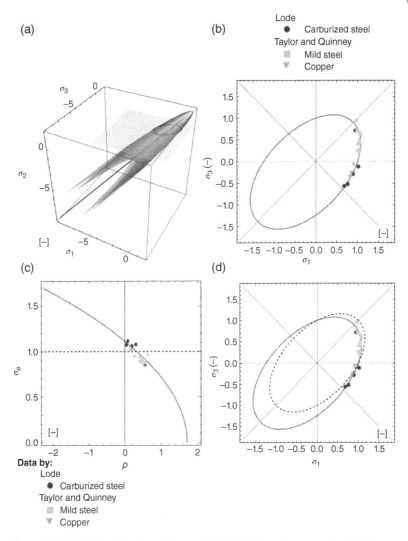

Figure 7.6 (a, b) Visualisation of the paraboloid yield surface according to the data of Lode (1926) and Taylor and Quinney (1931), collected by Theocaris (1995): (a) in the three-dimensional space of principal stresses, (b) in the plane $\sigma_2 = 0$. (c, d) Visualisation of the paraboloid yield surface according to the data of Lode (1926) and Taylor and Quinney (1931), collected by Theocaris (1995): (c) in the coordinate system (σ_e, p), (d) comparison with the Huber–Mises criterion in the plane $\sigma_2 = 0$. *Source:* Copyright by Teresa Frąś.

strictly quadratic hypothesis of material effort for anisotropic solids. The new concept of energy orthogonal decomposition of stress state was the subject of the study. The quadratic limit criteria cannot account for the asymmetry of the elastic range. Therefore, Burzyński (1928, 2009)

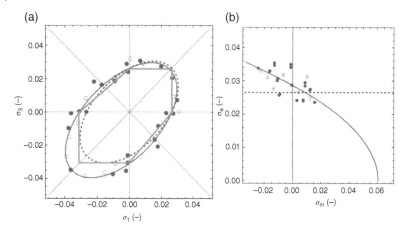

Figure 7.7 (a, b) Visualisation of the Burzyński paraboloidal failure surface according to the data collected by Lund and Schuh (2003) and Schuh and Lund (2003) for metallic glass and comparison with the Huber–Mises yield condition and Coulomb–Mohr criterion for the offset strain 0.0005: (a) in the plane $\sigma_2 = 0$, (b) in the coordinate system (σ_e, p). *Source:* Copyright by Teresa Frąś.

considered only a particular class of linear elastic materials for which the decomposition of the stored elastic energy density Φ into the sum of energy of volume change Φ_V and energy of distortion Φ_f is possible in some cases of anisotropy:

$$\Phi = \frac{1}{2} A_\sigma \cdot A_\varepsilon + \frac{1}{2} D_\sigma \cdot D_\varepsilon \tag{7.13}$$

where A_σ, A_ε and D_σ, D_ε denote the spherical and deviatoric parts of stress and strain tensors, respectively. Such a class of linearly elastic anisotropic solids is called volumetrically isotropic. It is worthy to note that the problem of volumetric-distortional decomposition, considered initially by Burzyński in 1928, was undertaken more recently by Ting (2001) and Federico (2010). The energy-based condition for a particular class of orthotropic materials exhibiting the elastic range's asymmetry is seen in (Pęcherski et al. 2020). This paper contains a historical account of the energetic hypothesis of material effort for anisotropic solids revealing the SDE. It is extended with the detailed discussion of the numerous works based on other methodology, as well.

For example, consider applying the elastic limit criterion derived in Pęcherski et al. (2020) compared to the experimental results for E335 steel (Frąś 2013) with a body cubic centred (BBC) microstructure. The results of quasi-static investigations of thin-walled tubes of E335 steel subjected to a

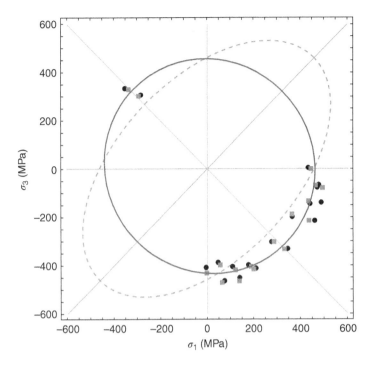

Figure 7.8 The limit curves describe the experimental data for E335 steel, the red line Burzyński yield condition, the dashed blue line Huber–Mises criterion, cf. Pęcherski et al. (2020). *Source:* Copyright by Teresa Frąś.

complex state of stress are used to determine the yield surface. In Figure 7.8, the limit yield curves obtained according to the smooth approximation of the Burzyński yield condition (dark grey line) and the classical Huber–Mises criterion (dashed light grey line), confronted with the experimental data, are presented. Let us observe that there is a remarkable difference between both criteria lines. The reason is that there is a difference between yield point in shear obtained experimentally (325 MPa) and calculated one due to theoretical relation $\tau_Y = \sigma_Y/\sqrt{3}$. The difference is almost 20% because the theoretical value equals 260 MPa.

References

Bridgman, P.W. (1947). The effect of hydrostatic pressure on the fracture of brittle substances. *J. Appl. Phys.* 18: 246–258.

Bridgman, P.W. (1952). *Studies in Large Plastic Flow and Fracture With Special Emphasis on the Effects of Hydrostatic Pressure*. New York: McGraw-Hill.

Burzyński, W. (2008). Theoretical foundations of the hypotheses of material effort. Włodzimierz Burzyński (1900–1970). *Eng. Trans.* 56: 189–225. English translation from the original edition in Polish: Teoretyczne podstawy hipotez wytężenia. *Czasopismo Techniczne.* 47: 1–41 Lwów, 1929.

Burzyński, W. (2009). Selected passages from Włodzimierz Burzyński's doctoral dissertation *Study on material effort hypotheses. Eng. Trans.* 57: 185–215. English translation from original edition in Polish: *Studium nad Hipotezami Wytężenia.* Akademia Nauk Technicznych, Lwów, 1928.

Christensen, R.M. (2004). A two-property yield, failure (fracture) criterion for homogeneous, isotropic materials. *J. Eng. Mater. Technol.* 126: 45–52.

Christensen, R.M. (2006). A comparative evaluation of three isotropic, two property failure theories. *J. Appl. Mech.* 73: 852–859.

Coffin, L.F. (1950). The flow and fracture of brittle materials. *J. Appl. Mech.* 17: 233–248.

Drucker, D.C. and Prager, W. (1952). Soil mechanics and plastic analysis for limit design. *Q. Appl. Math.* 10: 157–165.

Federico, S. (2010). Volumetric-distortional decomposition of deformation and elasticity tensor. *Math. Mech. Solids* 15: 672–690.

Finnis, M. (2003). *Interatomic Forces in Condensed Matter.* Oxford: Oxford University Press.

Frąś, T. (2013). Modelling of plastic yield surface of materials accounting for initial anisotropy and strength differential effect on the basis of experimental and numerical simulation. PhD thesis. Metz & Krakow: Université de Lorraine & AGH University of Science and Technology.

Frąś, T. and Pęcherski, R.B. (2010). Applications of the Burzyński hypothesis of material effort for isotropic solids. *Mech. Control* 29: 45–50.

Frąś, T., Kowalewski, Z.L., Pęcherski, R.B., and Rusinek, A. (2010). Application of Burzyński failure criteria. Part I. Isotropic materials with asymmetry of elastic range. *Eng. Trans.* 58: 2–13.

Frąś, T., Nowak, Z., Perzyna, P., and Pęcherski, R.B. (2011). Identification of the model describing viscoplastic behaviour of high strength metals. *Inverse Prob. Sci. Eng.* 19: 17–30.

Freudenthal, A.M. (1950). *The Inelastic Behavior of Engineering Materials and Structures.* New York: Wiley.

Gilman, J. (2003). *Electronic Basis of Strength of Materials.* Cambridge: Cambridge University Press.

Grassi, R. and Cornet, I. (1949). Fracture of grey cast-iron tubes under biaxial stresses. *J. Appl. Mech.* 17: 178–182.

Jirasek, M. and Bazant, Z.P. (2002). *Inelastic Analysis of Structures.* Chichester: Wiley.

Lode, W. (1926). Versuch über den Einfluß der mittleren Hauptspannung auf das Fließen der Metalle Eisen, Kupfer und Nickel. *Z. Phys.* 36: 13–39.

Lund, A.C. and Schuh, C.A. (2003). Yield surface of a simulated metallic glass. *Acta Mater.* 51: 5399–5411.

Mises, R.V. (1928). Mechanik der plastischen Formanderung von Kristallen. *ZAMM* 8: 161–185.

Mishin, M.J., Mehl, D.A., Papaconstantopoulos, A.F. et al. (2001). Structural stability and lattice defects in copper: Ab initio, tight-binding, and embedded-atom calculations. *Phys. Rev. B* 63: 224106.

Nalepka, K. (2012a). Efficient approach to metal/metal oxide interfaces within variable charge model. *Eur. Phys. J. B* 85: 45. https://doi.org/10.1140/epjb/e2011-10839-1.

Nalepka, K. (2012b). Symmetry-based approach to parametrization of embedded – atom – method interatomic potentials. *Comput. Mater. Sci.* 56: 100–107.

Nalepka, K. (2013). Material symmetry: a key to specification of interatomic potentials. *Bull. Pol. Acad. Sci.: Tech. Sci.* 61: 1–10.

Nalepka, K. and Pęcherski, R.B. (2009). Modelling of the interatomic interaction in the copper crystal applied in the structure (111)Cu‖(0001) Al_2O_3. *Arch. Metall. Mater.* 54: 511–522.

Nalepka, K., Sztwiertnia, K., Nalepka, P., and Pęcherski, R.B. (2015). The strength analysis of Cu/α-Al_2O_3 interfaces as a key for rational composite design. *Arch. Metall. Mater.* 60: 1953–1956.

Nardin, A., Zawarise, G., and Schrefler, B.A. (2003). Modelling of cutting tool soil interaction. Part. I. Contact behaviour. *Comput. Mech.* 31: 327–339.

Nowak, Z., Perzyna, P., and Pęcherski, R.B. (2007). Description of viscoplastic flow accounting for shear banding. *Arch. Metall. Mater.* 52: 217–222.

Olszak, W. and Ostrowska-Maciejewska, J. (1985). The plastic potential in the theory of anisotropic elastic-plastic solids. *Eng. Fract. Mech.* 21: 625–632.

Pęcherski, R.B. (1983). Relation of microscopic observations to constitutive modelling of advanced deformations and fracture initiation of viscoplastic materials. *Arch. Mech.* 35: 257–277.

Pęcherski, R.B. (1997). Macroscopic measure of the rate of deformation produced by micro-shear banding. *Arch. Mech.* 49: 385–401.

Pęcherski, R.B. (1998a). Macroscopic effects of micro-shear banding in plasticity of metals. *Acta Mech.* 131: 203–224.

Pęcherski, R.B. (1998b). Macromechanical description of micro—shear banding. *Studies in Applied Mechanics* 46: 203–222, Elsevier.

Pęcherski, R.B. (2008). Burzyński yield condition vis-à-vis the related studies reported in the literature. *Eng. Trans.* 56: 311–324.

Pęcherski, R.B., Szeptyński, P., and Nowak, M. (2011). An extension of Burzyński hypothesis of material effort accounting for the third invariant of stress tensor. *Arch. Metall. Mater.* 56: 503–508.

Pęcherski, R.B., Nalepka, K., Frąś, T., and Nowak, M. (2014). Inelastic flow and failure of metallic solids. Material effort: study across scales. In: *Constitutive*

Relations under Impact Loadings. CISM International Centre for Mechanical Sciences (ed. T. Łodygowski and A. Rusinek), 245–285. CISM Udine. https://doi.org/10.1007/978-3-7068-2_6.

Pęcherski, R.B., Rusinek, A., Fras, T. et al. (2020). Energy-based yield condition for orthotropic materials exhibiting asymmetry of elastic range. *Arch. Metall. Mater.* 65: 771–778.

Phillips, R. (2001). *Crystals, Defects and Microstructures. Modelling Across Scales.* Cambridge: Cambridge University Press.

Pisarenko, G.S. and Lebedev, A.A. (1973). *The Deformation and Strength of Materials under Complex State of Stress.* Kiev: Naukova Dumka (in Russian).

Richmond, O. and Spitzig, W.A. (1980). Pressure dependence and dilatancy of plastic flow. *International Union for Theoretical and Applied Mechanics Conference Proceedings*, pp. 377–386.

Rose, J.H., Smith, J.R., Guinea, F., and Ferante, J. (1984). Universal features of the equation of state of metals. *Phys. Rev.* B29: 2963.

Rychlewski, J. (1984). Elastic energy decomposition and limit criteria. *Eng. Trans.* 59: 31–63. 2011, English translation of the original paper in Russian from Advances in Mechanics (Uspekhi Mekhaniki), 7: 51–80. 1984.

Schuh, C. and Lund, A.C. (2003). Atomistic basis for the plastic yield criterion of metallic glass. *Nat. Mater.* 2: 449–452.

Sendeckyj, G.P. (1972). Empirical strength theories. Testing for production of material performance in structures and components. *ASTM STP* 515: 171–179.

Spitzig, W.A., Sober, R.J., and Richmond, O. (1976). The effect of hydrostatic pressure on the deformation behavior of maraging and HY-80 steels and its implications for plasticity theory. *Metall. Trans.* A 7A: 1703–1710.

Taylor, G. and Quinney, H. (1931). The plastic distortion of metals. *Philos. Trans. R. Soc.* A230: 323–362.

Theocaris, P.S. (1995). Failure criteria for isotropic bodies revisited. *Eng. Fract. Mech.* 51: 239–264.

Ting, T.C.T. (2001). Can a linear anisotropic elastic material have a uniform contraction under a uniform pressure? *Math. Mech. Solids* 6: 235–243.

Vadillo, G., Fernandez-Sáes, J., and Pęcherski, R.B. (2011). Some applications of Burzyński yield condition in metal plasticity. *Mater. Des.* 32: 628–635.

Wilson, C.D. (2002). A critical reexamination of classical metal plasticity. *J. Appl. Mech.* 69: 63–68.

Wolfram Mathematica, 6.0 Copyright (1988–2007). Wolfram Research Inc. Oxfordshire, UK.

Yoshikawa, T., Tokuda, M., Inaba, T. et al. (2008). Plastic deformation of AZ31 magnesium alloy under various temperature conditions. *J. Soc. Mater. Sci. Jpn.* 57 (7): 688–695.

Yu, M.-H. (2004). *Unified Strength Theory and its Applications*. Berlin, Heidelberg: Springer-Verlag.

Yu, M.-H., Qiang, G.-W., and Zhang, Y.-Q. (2006). *Generalized Plasticity*. Berlin, Heidelberg: Springer-Verlag.

8

Constitutive Description of Viscoplasticity Accounting for Shear Banding

Based on the discussion in Section 6.4, it seems that one could directly introduce an elastoplasticity model with two limit surfaces. The study of the plastic flow processes with an external surface accounting for the onset of shear banding, serving as the visualisation of 'extremal surface' and the internal surface relating to the elastic range and back stress anisotropy with nonlinear evolution of kinematic hardening appeared in Pęcherski (1995, 1996). In the work of Pęcherski (1998), a more extended discussion and comprehensive analysis are also provided. Therefore, we refer first to the details in the cited papers to show the constitutive description background and attract potential readers' attention. After that, the further study of viscoplastic flow processes accounting for shear banding becomes the new subject of more detailed presentation and discussion.

8.1 The Model of Plastic Flow with Nonlinear Development of Kinematic Hardening

In the 1970s and 1980s, computational modelling for metal shaping and ductile fracture with related strain localisation phenomena produced an increasing demand for an adequate constitutive description of the inelastic behaviour of metallic solids. An account for deformation-induced anisotropy, e.g. kinematic hardening, appeared particularly demanding. The necessity of introducing tensorial internal variables, particularly the kinematic hardening parameter (back stress), relates to the formulation of objective rate-type constitutive equations. Unfortunately, the early analysis revealed that applying the known Zaremba–Jaumann time derivative can lead to a non-adequate prediction of the material reaction. Then

Viscoplastic Flow in Solids Produced by Shear Banding, First Edition. Ryszard B. Pęcherski.
© 2022 John Wiley & Sons Ltd. Published 2022 by John Wiley & Sons Ltd.

significantly, the finite shear strain with large rotations of the back stress tensor principal axes plays the dominant role. Dienes (1972) demonstrated that the solution of the problem of finite simple shear of hypoelastic material with the Zaremba–Jaumann rate predicts stresses that oscillate as the shear strain increases. Lehmann (1972) and a decade later (Nagtegaal and De Jong 1982) found the unwanted oscillatory stresses generated by finite simple shear in plastic materials with kinematic hardening and Zaremba–Jaumann stress rate. Lee et al. (1983) and Onat (1984) discussed certain modifications of the objective rate of stress. However, the predicted results did not stand the time and experimental confrontation test. A better situation appeared with the use of the Mandel concept of director vectors by Loret (1983), Fressengeas and Molinari (1983), and Dafalias (1983). The comprehensive discussion and analytic solutions of the finite simple shear and numerical calculations of simple shear traction problems provided in Paulun and Pęcherski (1985, 1987) led to the conclusion that the known in the literature and cited extensively in these two works modifications of the relation for plastic spin led to non-adequate predictions of material behaviour, cf. Paulun and Pęcherski (1985) Figs. 6 and 7, as well as Paulun and Pęcherski (1987), Fig. 5. The main reason for these deficiencies is the use of the Prager-type model of the back stress evolution

$$\overset{\circ}{\alpha} = 2h_\alpha D^p \tag{8.1}$$

with the Zaremba–Jaumann time derivative, of the back stress tensor α, responsible for kinematic hardening

$$\overset{\circ}{\alpha} = \dot{\alpha} - W^e \alpha + \alpha W^e, \tag{8.2}$$

where the relation for the substructure elastic spin

$$\mathbf{W}^e = \mathbf{W} - \mathbf{W}^p \tag{8.3}$$

demands the additional constitutive equation for the plastic spin $\mathbf{W}^p$, Paulun and Pęcherski (1985) and Pęcherski (1988):

$$\mathbf{W}^p = \eta \left(\alpha D^p - D^p \alpha \right), \tag{8.4}$$

while η is a scalar function of the isotropic invariants of the kinematic hardening parameter α instead of a constant coefficient. However, it is visible in the literature on finite inelastic deformations accounting for strain-induced anisotropy that the assumption about the misleading constancy of the multiplier η in (8.4) is prevalent. Unfortunately, this might lead the readers to describe and predict simulated processes of finite inelastic deformations incorrectly.

The more refined specification of the influence function η comes from the deeper reflection on the relation for plastic spin transpiring from a physical interpretation of the relative motion of a continuum vis-à-vis the material substructure (crystalline lattice), cf. Paulun and Pęcherski (1992a,b). Let us assume the small elastic and finite plastic deformation model with Huber–Mises yield condition and combined nonlinear hardening rule. The constitutive equations at the yield point take the form:

$$f(S,\alpha,k) = \frac{1}{2}(S-\alpha):(S-\alpha) - k^2(\gamma_{eq}) = 0 \tag{8.5}$$

$$D = D^e + D^p = \left(L^{-1} + \frac{\alpha}{2h}\mu_f \otimes \mu_f\right):\tau, \tag{8.6}$$

while

$$\mu_f = \frac{1}{\sqrt{2}}\frac{S-\alpha}{k} \tag{8.7}$$

and the nonlinear kinematic hardening law reads

$$\overset{\circ}{\alpha} = 2h_\alpha D^p - c_r \dot{\gamma}_{eq} , \tag{8.8}$$

$$\dot{\gamma}_{eq} = \left(D^p : D^p\right)^{1/2}, \tag{8.9}$$

where L^{-1} is the fourth-order tensor of elastic moduli, S represents the deviator of the Kirchhoff stress τ, the parameter k corresponds to the 'size' of the yield surface. The material constant c_r relates to saturation of the back stress effect while accumulation of plastic strain is growing. The nonlinear form of the kinematic hardening evolution Eq. (8.8) can be traced back to the work of Armstrong and Frederick (1966), who proposed a similar hardening hypothesis with evanescent memory for small deformations. The combined plastic hardening modulus h reads:

$$h = h_i + h_\alpha, \tag{8.10}$$

where isotropic hardening modulus is determined as follows

$$h_i = \frac{\partial k}{\partial \gamma_{eq}} \tag{8.11}$$

and the nonlinear kinematic hardening modulus takes the form:

$$h_\alpha = h'_\alpha - \frac{1}{\sqrt{2}} c_r \mu_f : \alpha. \tag{8.12}$$

The scalar multiplier α in (8.6) fulfils the conditions $\alpha = 1$ if $\mu_f : \overset{\circ}{\tau} \geq 0$ and $\alpha = 0$ if $\mu_f : \overset{\circ}{\tau} < 0$.

Finally, the additional constitutive equation was derived for the plastic spin $\mathbf{W}^p$ (8.4), while Pęcherski (1995):

$$\eta = \frac{1}{\left(h'^2_\alpha + \frac{\sqrt{3}}{3} \alpha_{eq} h'_\alpha \right)^{1/2}} \tag{8.13}$$

and

$$\alpha_{eq} = \left(\frac{1}{2} \alpha : \alpha \right)^{1/2}. \tag{8.14}$$

The detailed discussion of the properties of relations for a plastic spin in Paulun and Pęcherski (1992a,b) reveals that the Eqs. (8.4) and (8.13) fulfil the continuity requirement, $lim(\eta\alpha) = 0$ as $\alpha \to 0$, and are applicable for the numerical modelling of the processes under monotonic as well as reverse loading conditions. The background of deriving the properly justified experimentally constitutive equation for the plastic spin (8.4) and (8.13), presented in Paulun and Pęcherski (1992b), is worth mentioning. As discussed in Pęcherski (1988), the relation (8.4) appears as an approximation of more general formula of the skew–symmetric constitutive tensor function of plastic spin discussed in Loret (1983) and Dafalias (1983). Also, the form of the scalar function (8.13) results from the analysis of the finite simple shear problem's solution (Paulun and Pęcherski 1992b).

For comparison, the experimental data of M.G. Stout presented in Harren et al. (1989) as private communication denoted in the mentioned above paper as Stout (1984), p. 446, personal communication are displayed in Figure 8.1. Curve 1 corresponds to the numerical simulation of the Swift effect in finite unconstrained shear problem with the representation of the scalar function η given in the relation (Pęcherski 1988):

$$\eta = \sqrt{\frac{3}{2} \frac{6\alpha_{eq}}{\left(h'^2_\alpha + 3\alpha_{eq}^2 \right)^{\frac{1}{2}}}}. \tag{8.15}$$

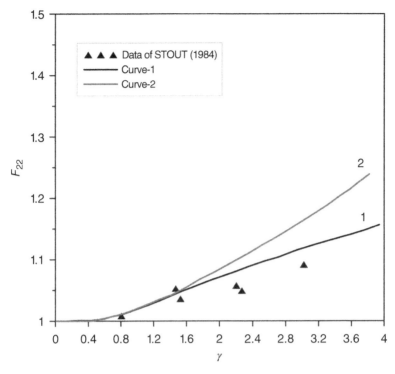

Figure 8.1 Swift effect for unconstrained shear as predicted numerically by the model with plastic spin expressed by (8.4) and two representations of the influence function η described in (8.13) – the curve 2 and in (8.15) – the curve 1. *Source:* Copyright by Ryszard Pęcherski.

Figure 8.1 shows an excerpt of the results displayed in Paulun and Pęcherski (1992b), cf. Fig. 3. Let us note that the relation (8.15) derived in Pęcherski (1988) found application in the numerical analysis of large plastic deformations with a strain-induced anisotropy, in particular, kinematic hardening and developing voids in the porous material (Lammering et al. 1990).

Let us observe that the nonlinear hardening rule implicitly introduces the second limit surface cf. Pęcherski (1995, 1998):

$$F\left(\mathbf{S}, k_{SB}\right) = \frac{1}{2}\left(\mathbf{S} : \mathbf{S}\right) - k_{SB}^2 = 0, \tag{8.16}$$

where k_{SB} is the 'size' of the limit surface (i.e. $k_{SB} = \frac{1}{\sqrt{2}} R$, while R is the radius of the external surface). According to the studies of Oliferuk et al. (1966), the following hypothesis transpires: the saturation of internal micro-stresses correlates with the massive formation of micro-shear bands, which means the initiation of shear banding. This phenomenon relates to a certain amount of plastic strain accumulated along a given deformation path.

In particular, Fig. 4 of Pęcherski (1998) displays the generic microscopic shear-banding limit approximated by the class of the external surfaces obtained for different loading paths, which relate the saturation of the back stress effect with micro-shear banding. One should note that systematic experimental investigations are necessary to identify the mentioned shape and evolution of 'extremal surface'. It seems an ambitious task to realise in the future.

8.2 The Perzyna Viscoplasticity Model Accounting for Shear Banding

The viscoplasticity theory proposed originally by Perzyna (1963), extended within a more general framework of rheology (Perzyna 1966) and with the background of thermodynamics of materials (Perzyna 1971), applies here in the case of the ultrafine-grained (ufg) and nanocrystalline metals (Nowak et al. 2007). The particular specification of the nonlinear excess stress function (Perzyna 1963), as the power law, takes the form:

$$\mathbf{D}^p = \frac{\sqrt{2}}{2}\dot{\gamma}_S\mu, \quad \mu = \frac{1}{\sqrt{2}k_s}\mathbf{S}, \tag{8.17}$$

$$\dot{\gamma}_s = \dot{\gamma}_0\left[\frac{J_2}{k_s} - 1\right]^{\frac{1}{D}} \; for\, J_2 - k_s > 0, J_2 = \sqrt{\frac{1}{2}\mathbf{S}:\mathbf{S}} \tag{8.18}$$

and

$$\dot{\gamma}_s = 0 \; for \; J_2 - k_s \leq 0, \tag{8.19}$$

where k_s is quasistatic yield limit controlled by crystallographic slips. Finally, according to the above Perzyna viscoplasticity model, accounting for the mechanism of shear banding yields:

$$\dot{\gamma} = \frac{\dot{\gamma}_0}{\left(1 - f_{SB}\right)}\left[\frac{J_2}{k_s} - 1\right]^{\frac{1}{D}} \; for\, J_2 - k_s > 0. \tag{8.20}$$

Inverting (8.20) gives the relation for the dynamic yield condition

$$J_2 = k_s^d = k_s \left\{ 1 + \left[\left(\frac{\dot{\gamma}}{\dot{\gamma}_0} (1 - f_{SB}) \right)^D \right] \right\}. \tag{8.21}$$

The relation for the dynamic yield shear strength k_d can be derived for the representative volume element (RVE), in which the mechanisms of crystallographic slip and shear banding are competitively operative. The derivation comes from the balance of the sum of the dissipation rate produced by each mechanism and the dissipation rate of the deformation in the whole RVE. It is under the assumption that the material reveals no work hardening. It also assumes that the value of the threshold stress of shear banding is negligible compared to the value of the dynamic yield condition in (8.21). As a result, the following relation is derived:

$$k_d = \left(1 - f_{SB} \right)\left(1 - f_{SB}^V \right) k_s^d, \tag{8.22}$$

where $f_{SB}^V = \dfrac{V_{SB}}{V}$ denotes the volume fraction of the region of RVE in which the shear-banding mechanism operates. As mentioned in the Preface, the afore-defined ratio corresponds, besides the *instantaneous shear-banding contribution function* f_{SB} to a *cumulative kind of shear banding* f_{SB}^V.

Three particular cases can take place:

1) In some instances related mainly to the deformation processes in the ufg and nanocrystalline metals, the activity of a cumulative kind of shear banding f_{SB}^V prevails, and therefore the assumption $f_{SB} \approx f_{SB}^V$ holds and the shape of the relation (8.22)

$$k_d = \left(1 - f_{SB}^V \right)^2 k_s^d \tag{8.23}$$

is justified cf. Nowak et al. (2007). Therefore, we have:

$$k_d = \left(1 - f_{SB}^V \right)^2 k_s \left\{ 1 + \left[\left(\frac{\dot{\gamma}}{\dot{\gamma}_0} (1 - f_{SB}) \right)^D \right] \right\}. \tag{8.24}$$

2) The other case corresponds to the different extreme situations. The dominant mechanism of shear banding appears in polycrystalline metallic solid deforming in multiple slip systems. The instantaneous shear-banding contribution function (6.29) or (6.32) for the symmetric shear-banding system plays a decisive role. In that case, the shape of the relation for dynamic yield condition transforms, cf. (8.22):

$$k_d = \left(1 - f_{SB}\right)^2 k_s \left\{ 1 + \left[\left(\frac{\dot{\gamma}}{\dot{\gamma}_0} (1 - f_{SB}) \right)^D \right] \right\} \quad (8.25)$$

3) The material with a hybrid structure – the dislocation-mediated crystallographic and partly isomorphic and possibly nanocrystalline with the obstructed mechanism of dislocation breading can be the study's subject due to minimal grain size that is lower than 100 nm. It also corresponds particularly to glassy metals with a developed partially crystalline phase. In such a case, the following relation for dynamic yield condition holds:

$$k_d = \left(1 - f_{SB}\right)\left(1 - f_{SB}^V\right) k_s \left\{ 1 + \left[\left(\frac{\dot{\gamma}}{\dot{\gamma}_0} (1 - f_{SB}) \right)^D \right] \right\}. \quad (8.26)$$

These three cases fulfil the possible applications in modelling viscoplastic flow accounting to or produced by shear banding.

8.3 Identification of the Viscoplasticity Model

To identify the description of inelastic deformation accounting for shear banding, consider the problem of constrained plane strain compression, which approximates the commonly applied in a laboratory – channel-die test (Bronkhorst et al. 1992; Anand et al. 1994), cf. the schematic picture in Fig. 2.2. Due to the slow quasi-static processes, the material's inelastic behaviour is described by applying the rate-independent elastic–plastic model with work hardening cf. Pęcherski and Korbel (2002). Nowak and Pęcherski (2002) present numerical calculations with the finite element program ABAQUS (ABAQUS/STANDARD 2001). The new algorithms developed in Nowak and Stachurski (2001, 2002) found application by that. The solution of the nonlinear regression problem using the method of global optimisation (Boender et al. 1982) appeared necessary with the automatic procedure of identifying the unknown scalar function describing the contribution of shear bands in plastic shear–strain rate.

The mentioned above papers of Bronkhorst et al. (1992) and Anand et al. (1994) present the experimental results of the channel-die test for samples of polycrystalline copper. The main subject of the study was the texture evolution and development of micro-shear bands. The matrix and experimental setting geometry is displayed in Figure 2.2. These are the following dimensions of the sample:

- the height corresponding to the direction of compression: $e_3 - 6.35$ mm
- the width corresponding to the direction of the free plastic flow: $e_2 - 9.53$ mm
- the length corresponding to the direction of matrix constraints: $e_1 - 14.73$ mm.

The application of Teflon plates covering the contact surfaces of the matrix and the sample reduces the usual effects of friction. Therefore, in numerical calculations, the assumption about the frictionless contact between the sample and the matrix is justified. The experimental studies included four series of compression tests reaching the logarithmic strain e_3 values of -0.21; -0.52; -1.0; and -1.54. The assumed compression strain rate equaled 0.001 s^{-1}. They investigated, with the use of the metallographic means, surfaces of deformed samples lead to the following observations (Anand et al. 1994), p. 234: '. . . at a true strain of 0.21, the grains are still reasonably equiaxed. Figure 10a shows the microstructure at a true strain of -052. At this stage the grains have been flattened, and localised band, which are $0.1 \div 0.5$ μm in thickness and inclined at $\sim\pm 30 \div 40°$ to the horizontal can be observed within individual grains. The initiation of such micro-shear bands occurs somewhere between true strain levels of -0.21 and -0.52. The intensity of these micro-shear bands continues to increase as deformation progresses, and by a strain level of -1.00, Fig. 11a, macro-shear bands which cross grain boundaries have formed'.

In the works of Bronkhorst et al. (1992) and Anand et al. (1994), the authors present numerical simulations of the similar process of constrained plane strain compression, approximated by plane strain state polycrystalline aggregate with the application of finite element program ABAQUS. The aggregate consists of many grains with assumed crystal lattice orientations, in which the three-dimensional activation of all potential slip systems is allowed. The viscoplastic flow law at the level of a single-slip system is active. A separate finite element corresponds to a single crystal or a part of it. In the constrained plane strain compression analysis, an aggregate of 400 grains is represented by 400 quadrilateral continuum plane strain four-node elements – CPE4. The frictionless process, i.e. the lack of friction between the sample's surface, matrix, and punch, appeared in the computations as an idealisation. The confrontation of the numerical calculations and experimental data is displayed in Figure 8.2. The measurement results represent the points taken from the plot of experimental data presented in Anand et al. (1994), p. 457 – Fig. 7. Let us observe that for strain values taken from the range 0.21–0.52, the metallographic observations reveal the

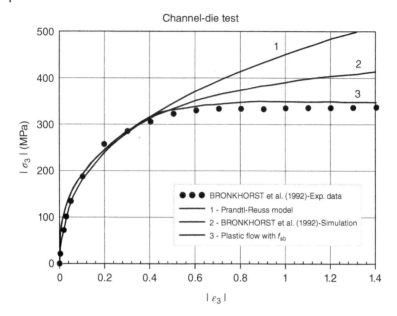

Figure 8.2 The plot of the absolute value of compression stress $|\sigma_3|$ as a function of logarithmic strain $|e_3|$ for different models of plastic flow versus the experimental results of the channel-die test: points represent experimental data of Bronkhorst et al. (1992), curve 1 – results of Prandtl–Reuss model for isotropic hardening, curve 2 – numerical simulation of the aggregate of grains according to Bronkhorst et al. (1992), curve 3 – results of plastic flow law model accounting for shear-banding contribution function f_{SB}. *Source:* Copyright by Ryszard Pęcherski.

development of micro-shear bands and the discrepancy between the determined points experimentally and the curve displaying the results of numerical computations becomes visible. The disparity increases in the increase of strain, reaching about 22% for the value of logarithmic strain $|e_3| = 1.4$. The observed inconsistency of numerical simulation and experimental observations relates to the development of micro-shear banding and the increase of the contribution of this new mechanism in the process of constrained compression in the channel-die test reported in Bronkhorst et al. (1992). It might be the reason for this discrepancy.

The ABAQUS/Standard uses the homogeneous plane strain compression to simulate the idealised frictionless channel-die test (Nowak and Pęcherski 2002). Plastic flow law accounting for the contribution of the symmetric system of micro-shear bands is used (Pęcherski and Korbel 2002), and the model derived is specified for infinitesimal elastic strains in the form of the following rate-type equation:

$$\overset{\circ}{\sigma}_{ij} = C_{ijkl} D_{kl}, C_{ijkl} = 2G\left(\delta_{ik}\delta_{jl} + \frac{v}{1+v}\delta_{ij}\delta_{kl} - \frac{1}{\alpha\sigma_Y^2}\sigma'_{ij}\sigma'_{kl}\right), \qquad (8.27)$$

where the symbol $\overset{\circ}{\sigma}_{ij}$ denotes the components of the Zaremba–Jauman derivative of the Cauchy stress tensor and σ'_{ij} are the components of the deviator of the Cauchy stress tensor on an orthonormal basis. The parameter α reads:

$$\alpha = \frac{2}{3}\left[1 + \frac{h(1-f_{SB})}{3G}\right], \qquad (8.28)$$

where h is plastic-hardening modulus. Let us observe that the contribution of shear banding $f_{SB} \in [0,1)$ is accounted in the scalar parameter α, affecting elastoplastic moduli and, at the same time, the stiffness matrix of the considered numerical scheme. The results of computations based on the classical model of elastic–plastic deformation with isotropic hardening (called in the literature Prandtl-Reuss model or J_2 theory) are represented in Figure 8.2 by curve 1. Observe that the large discrepancy between curve one and the experimental data is visible. For the strain $|e_3| = 1.4$ the difference reaches about 53%. The numerical simulation results for the elastoplasticity model defined by (8.27) are represented by curve 3. The calculations are performed for such a form of the scalar function $f_{SB} = F_{SB}(\varepsilon^p)$, which describes the dependence of shear-banding contribution on equivalent plastic strain ε^p assuring the possibly close fitting of curve 3 concerning the experimental points. The possibility provided by the program ABAQUS/Standard was used, which enabled modification of material description by the user procedure (UMAT). In this procedure, the simple version of the known radial return algorithm is implemented to integrate the elastoplasticity flow rule with the Huber–Mises yield condition and isotropic hardening. The calculations were performed for the approximation of hardening curve taken from the free compression test of Cu polycrystalline samples presented in Bronkhorst et al. (1992), p. 450, Fig. 2:

$$\sigma = \sigma_Y\left(\frac{E}{\sigma_Y}\varepsilon^p\right)^{\frac{1}{m}}, \qquad (8.29)$$

where the yield limit $\sigma_Y = 0.02$ GPa, the Young modulus $E = 126.0$ GPa, $m = 2.93$. The shape of the sought shear-banding contribution function $f_{SB} = F_{SB}(\varepsilon^p)$, which provides the required fitting of experimental points with the assumed 3% error, is given in Figure 8.2. The function is taken as a logistic (sigmoidal) function of the form

$$F_{SB}\left(\varepsilon^p\right) = \frac{f_{SB0}}{1 + \exp\left(a - b|\varepsilon_3|\right)} \tag{8.30}$$

Such an identification method relies on an intuitive selection of the form of the sought function and fitting of relevant constants using the series of repeated calculations of the process of homogeneous constrained plane strain compression and comparison of the computed values of compression stress σ_3, displayed in Figure 8.2 – curve 3, with the experimental data points presented in Bronkhorst et al. (1992). An automatic iterative procedure made the identification more efficient and independent of intuitive guesses. The specific starting value of the shear-banding contribution function takes the form, e.g. $f_{SB}^{start} = 0.5$, and the series of calculations is executed for plane strain compression. On each step for a given strain increment $|\Delta\varepsilon_3|$, one can check if the stress values σ_3 calculated for different values of $f_{SB} \in [0, 1)$, lay sufficiently near, with an assumed error, to the values taken from the curve representing the experimental data (cf. Figure 8.2). Figure 8.3 results of the automatic identification procedure correspond to the assumed error equal to 5%. It appears that an attempt of diminishing this value leads to a rapid increase in the number of iterations and the resulting cost of computations. The new algorithms developed in Nowak and Stachurski (2001, 2002) were implemented in the presented automatic identification procedure. The algorithms came from nonlinear regression problems using the global optimisation method (Boender et al. 1982).

8.4 The Crystal Plasticity Modelling of Deformation Processes in Metals Accounting for Shear Banding

The works of Kowalczyk-Gajewska et al. (2005) and Kowalczyk-Gajewska and Pęcherski (2009) deal with the description of polycrystalline metals deformation produced by two competing mechanisms: crystallographic slips and micro-shear bands formation. The study is based on the micromechanical rigid-plastic model for a single grain. The derived velocity gradient (5.23) splits into two parts connected with crystallographic slip and micro-shear banding. For a crystallographic slip, the regularised Schmid law of Gambin (1991, 1992) finds application. For the shear-banding mechanism,

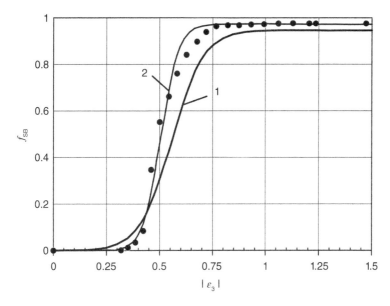

Figure 8.3 Comparison of the results of simple identification with the postulated function F_{SB}, (7.4), and fitted constants f_{SB0} = 0.95, a = 7.5, b = 13.6 – curve 1, with the results of automatic iterative identification – filled points, and its approximation – curve 2 with the use of the function (7.4) and the specified constants f_{SB0} = 0.975, a = 12.5, b = 25.0. *Source:* Copyright by Ryszard Pęcherski.

the above-discussed model accounts for the rate of plastic deformation according to the *instantaneous shear banding contribution function* f_{SB}. Different constitutive equations for the plastic spin due to two considered mechanisms of plastic flow are in use. The considered model applies for the simulation of crystallographic texture evolution in the polycrystalline element cf. Kowalczyk and Gambin (2004). One of the crucial achievements obtained by Kowalczyk–Gajewska and published in the joint papers (Kowalczyk-Gajewska et al. 2005; Kowalczyk-Gajewska and Pęcherski 2009) is the development of the description and numerical simulations of the essential influence of shear banding on texture development. One expects that adding the effects of shear banding stabilises texture evolution earlier than in pure crystallographic slips. Also, the confirmation of experimental observations of the prediction of texture phenomena is visible, cf. the cited papers of Kalidindi and Anand (1994) and Anand (2004) as well as Paul et al. (1996, 2003) and Stalony-Dobrzanski and Bochniak (2005).

The results mentioned above in Kowalczyk-Gajewska et al. (2005) and Kowalczyk-Gajewska and Pęcherski (2009) also contain a hypothesis delivering the answer for the question: what is the effect of the change of loading path on the shape of shear-banding contribution function f_{SB}? The answer takes the following form. Let the relation define such a function with the use of the given formula:

$$f_{SB}\left(\xi\right) = \frac{f_{SB0}}{1 + \exp\left(a - b\xi\right)},$$

(8.31)

where the symbol ξ denotes an invariant:

$$\xi = \sqrt{\frac{3}{2}}\,\varepsilon_{eq}^{p}\left(1 - \alpha\left|cos3\theta\right|\right), 0 \leq \alpha \leq 1.$$

(8.32)

The invariant ξ depends on the cumulated value of equivalent plastic strain and the rate of plastic deformation:

$$\dot{\varepsilon}_{eq}^{p} = \sqrt{\frac{2}{3}\mathbf{D}^{p} : \mathbf{D}^{p}},$$

(8.33)

and

$$\left|cos\theta\right| = 3\sqrt{6}\left|det\left(\mathbf{D}^{p} / \left\|\mathbf{D}^{p}\right\|\right)\right|.$$

(8.34)

Note that according to Figure 8.4 hypothesis which describes the effect of the change of loading path on the shape of shear-banding contribution function, the rapid increase of the instantaneous contribution function f_{SB} appears when the scheme of the proportional strain path is changing.

In Kowalczyk-Gajewska et al. (2005) and Kowalczyk-Gajewska and Pęcherski (2009) more detailed discussion on the results in Figures 8.4 and 8.5 are provided. Let us note that, in reality, strains vary within the aggregate and a single grain. The discussed model of a single grain is applicable in any micro–macro procedure. For example, using the self-consistent scheme, one may account for the heterogeneity of deformation over the grains, using the finite element approach for a polycrystalline aggregate. Anyway, the Taylor model has the virtue of simplicity and may give the results of a first-order approximation and agreement with experimental data. However, one would like to stress that the presented model only reduces the material-hardening rate produced by shear banding and the mentioned influence on the texture image, despite the assurance of its geometry and space distribution. The material parameters are identified using the experimental data of polycrystal-line copper reported in Kalidindi and Anand (1994) and Anand (2004). Figure 8.5 displays that in the case of simple compression, the shear banding contributes negligibly to the rate of plastic deformation.

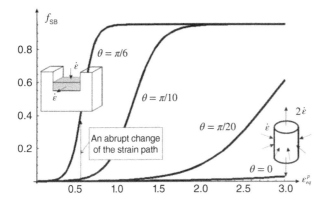

Figure 8.4 The variation of the instantaneous shear-banding contribution function f_{SB} in the course of different proportional strain paths determined by the angle θ. The two processes are specified: the uniaxial unconstrained tension, $\theta = \dfrac{\pi}{3}$ and the constrained plane strain (channel die test), $\theta = \dfrac{\pi}{6}$. *Source:* Copyright by Katarzyna Kowalczyk-Gajewska.

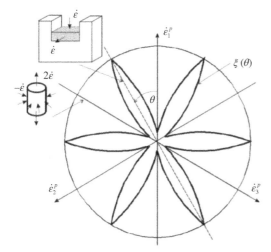

Figure 8.5 The variation of the invariant ξ for the prescribed constant value of the equivalent plastic strain in the deviatoric plane of the rate of plastic deformation tensor $\mathbf{D}^p$ for the proportional deformation path. *Source:* Copyright by Katarzyna Kowalczyk-Gajewska.

Let us note that the display in Figure 8.7 differs essentially from Figure 7 (Kowalczyk-Gajewska and Pęcherski 2009), containing more detailed information about the contribution of micro-shear bands. In particular, the shear banding contributing to plastic spin produces the deviation of the

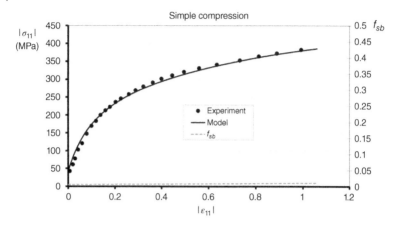

Figure 8.6 Axial stress versus logarithmic axial strain in simple compression compared with the experimental data of Anand (2004) revealing the negligible instantaneous contribution of shear banding f_{SB}, cf. Kowalczyk-Gajewska and Pęcherski (2009), Fig. 3. *Source:* Copyright by Katarzyna Kowalczyk-Gajewska.

stress–strain curve from the experimental points making the impediment of texture development. A similar effect appears in plane strain compression, cf. Fig. 6 (Kowalczyk-Gajewska and Pęcherski 2009), although it is not much visible. Anyway, Figures 8.6 and 8.7 as well as Figure 8.8 provide piece of essential information on the effect of shear-banding contribution purposely.

8.5 Viscoplastic Deformation of Nanocrystalline Metals

The subject of the study relates to ufg and nanocrystalline metals (nc metals). Experimental investigations discussed, e.g. by Meyers et al. (2006) and Jia et al. (2003), reveal that nanocrystalline materials exhibit very high yield strength, presumably due to the grain-boundary strengthening known as the Hall–Petch effect. It is generally faithful to the grain sizes, which are not less than one-tenth of a nanometer. The diminishing grain size leads to the situation that dislocation controlled mechanisms of inelastic deformation become hindered, and new dislocation-less mechanisms are activated. Then different kinds of grain boundary accommodation and shear banding (Meyers et al. 2006; Jia et al. 2003; Cheng et al. 2005) are reported. Experimental investigations of the behaviour of such materials under quasi-static as well as dynamic loading conditions related to microscopic observations show that in many cases, the dominant mechanism of plastic strain is multiscale development of shear deformation modes – called shear banding.

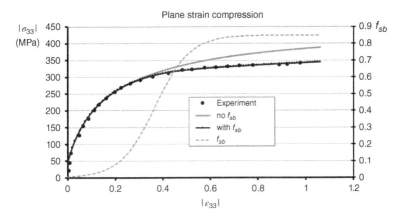

Figure 8.7 Axial stress versus logarithmic axial strain for plane strain compression compared with the experimental data of Anand (2004), revealing the visible increase of the instantaneous contribution of shear banding f_{SB}. The display of essential deviation of the stress–strain curve from the experimental points by neglecting the shear-banding contribution. *Source:* Copyright by Katarzyna Kowalczyk-Gajewska.

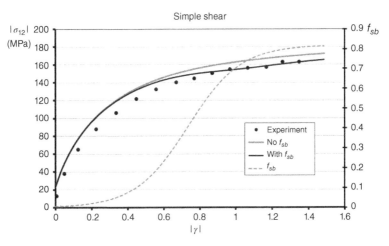

Figure 8.8 Shear stress versus amount of shear in simple shear compared with the experimental data of Anand (2004), revealing the moderate increase of the instantaneous contribution of shear banding f_{SB} and the visible deviation of the stress–strain curve from the experimental points by neglecting the shear-banding contribution. *Source:* Copyright by Katarzyna Kowalczyk-Gajewska.

Regarding nc-Fe Jia et al. (2003), for smaller grain sizes ($d < 300\,\text{nm}$), shear band development occurs immediately after the onset of plastic flow. Significant strain-rate dependence of the flow stress, particularly at high strain rates, was also emphasized. Kim et al. (2001) highlight grain boundary mechanisms and a Taylor-type phase mixture model used to calculate stress for describing the deformation of nanocrystalline metallic solids. In Wang and Yang (2004), another constitutive description was discussed, in which cooperative grain boundary mechanisms play a crucial role. The primary assumption is that plastic deformation accommodates grain boundaries, and the insertion and rotation processes are considered suitable deformation mechanisms. An overview of earlier attempts in accounting for grain boundary mechanisms is also provided.

On the other hand, viscoplastic models for the yield strength of nanocrystalline materials are derived in Lebensohn et al. (2006), Wei et al. (2006), and Wei and Anand (2007) emphasise that when the grain dimensions approach nanometer sizes, the volume fractions of grain boundaries increase the grain boundary regions and play an active role in accomodating deformations. The grain boundary accommodation may involve a local

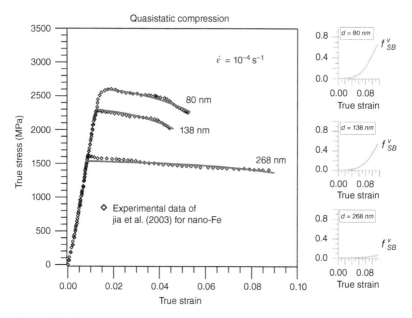

Figure 8.9 True stress-true strain curves for quasi-static compression tests for polycrystalline iron, mean grain diameters: 80, 138, and 268 nm. Continuous lines represent curves obtained from the viscoplasticity model accounting for shear banding. The symbols ◊ correspond to the experimental data for iron of purity 99.9% presented in Jia et al. (2003). On the right-hand side, the plots system of the volumetric contribution of shear banding f_{SB}^V versus true strain displays for the three-grain diameters: 80, 138, and 268 nm. *Source:* Copyright by Ryszard Pęcherski.

shuffling of atoms, grain boundary sliding, and different diffusive processes. Also, pressure and grain size dependency of yield strength were discussed. The works of Wei and Anand (2007) and Wei et al. (2006) deal with theoretical description and computational analysis of inelastic deformation of powder-processed nc-metals. The detailed analysis of micromechanical mechanisms controlling the inelastic behaviour can give physical motivation for the constitutive description based on two viscoplastic flow mechanisms: dislocation controlled slip and shear banding.

Nowak et al. (2007) and Frąś et al. (2011) consider the identification and computational simulations of quasi-static and dynamic compressive loading processes with the use of the viscoplasticity model accounting for shear banding for ufg and nc-metals. The results are displayed in Figures 8.9 and 8.10.

The above-discussed description of the viscoplastic behaviour of ufg and nc-metals can be comprehensive if one accounts for a more adequate yield

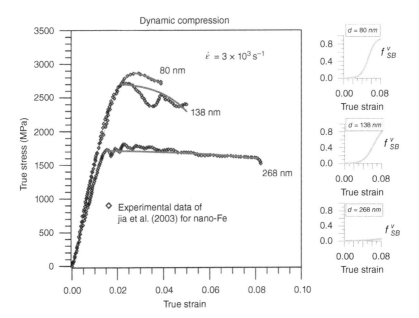

Figure 8.10 True stress–true strain curves for dynamic compression tests for polycrystalline iron, mean grain diameters: 80, 138, and 268 nm. Continuous lines represent curves obtained from the viscoplasticity model accounting for shear banding. The symbols ◊ correspond to the experimental data of dynamic compression tests for iron of purity 99.9% presented in Jia et al. (2003). On the right-hand side, the plots system of the volumetric contribution of shear banding f_{SB}^V versus true strain is displayed for the three-grain diameters: 80, 138, and 268 nm. *Source:* Copyright by Ryszard Pęcherski.

criterion. In the case of associated flow law, it also provides the appropriate potential function. According to the discussion in Chapter 7 such yield conditions may take in the principal stress axes the shape of rotationally symmetric paraboloid for isotropic materials and ellipsoidal paraboloid for materials possessing orthotropic symmetry, cf. also Frąś et al. (2011). The remarks on further applications of the relations describing the viscoplastic flow produced by shear banding are seen in Chapter 9.

References

ABAQUS Standard (2001). *Reference Manuals*. Providence: Hibbitt, Karlsson & Sorensen, Inc.

Anand, L. (2004). Single-crystal elasto-viscoplasticity: application to texture evolution in polycrystalline metals at large strain. *Comput. Methods Appl. Mech. Eng.* 193: 5359–5383.

Armstrong, P.J. and Frederick, C.O. (1966). A Mathematical Representation of the Multiaxial Bauschinger Effect. G.E.G.B. Report *RD/B/N 731*.

Boender, C.G., Rinnoykan, A.H.G., Strougie, L., and Timmer, G.T. (1982). A stochastic method for global optimization. *Math. Program.* 22: 125–140.

Bronkhorst, C.A., Kalidindi, S.T., and Anand, L. (1992). Polycrystalline plasticity and the evolution of crystallographic texture in FCC metals. *Philos. Trans. R. Soc. London, Ser. A* 341: 443–477.

Cheng, S., Ma, E., Wang, Y.M. et al. (2005). Tensile properties of in situ consolidated nanostructured Cu. *Acta Mater.* 53: 1521–1533.

Dafalias, Y.F. (1983). Corotational rates for kinematic hardening at large plastic deformations. *J. Appl. Mech.* 50: 561–565.

Dienes, K. (1972). On the analysis and stress rate in deforming bodies. *Acta Mech.* 32: 217–232.

Frąś, T., Nowak, Z., Perzyna, P., and Pęcherski, R.B. (2011). Identification of the model describing viscoplastic behaviour of high strength metals. *Inverse Problems Sci. Eng.* 19: 17–30.

Fressengeas, C. and Molinari, A. (1983). Représentation du comportement plastique anisotrope aux grandes déformations. *Arch. Mech.* 36: 483–498.

Gambin, W. (1991). Crystal plasticity based on yield surface with rounded-off corners. *ZAMM* **71**: T265–T268.

Gambin, W. (1992). Refined analysis of elastic-plastic crystals. *Int. J. Solids Struct.* **29**: 2013–2021.

Harren, S., Lowe, T.C., Asaro, R.J., and Needleman, A. (1989). Analysis of large-strain shear in rate-dependent FCC polycrystals: correlation of micro and macromechanics. *Philos. Trans. R. Soc. London, Ser. A* 328: 443–500.

Jia, D., Ramesh, K.T., and Ma, E. (2003). Effects of nanocrystalline and ultrafine grain sizes on constitutive behavior and shear bands in iron. *Acta Metall.* 51: 3495–3509.

Kalidindi, S.R. and Anand, L. (1994). Macroscopic shape change and evolution of crystallographic texture in pre-textured fcc metals. *J. Mech. Phys. Solids.* 42: 459–490.

Kowalczyk, K. and Gambin, W. (2004). Model of plastic anisotropy evolution with texture-dependent yield surface. *Int. J. Plasticity.* 20: 19–54.

Kowalczyk-Gajewska, K. and Pęcherski, R.B. (2009). Phenomenological description of the effects of micro-shear banding in micromechanical modelling of policrystal plasticity. *Arch. Metall. Mat.* 54: 1145–1156.

Lammering, R., Pęcherski, R.B., and Stein, E. (1990). Theoretical and computational aspects of large plastic deformations involving strain-induced anisotropy and developing voids. *Arch. Mech.* 42: 347–375.

Lebensohn, R.A., Bringa, E.M., and Caro, A. (2006). A viscoplastic micromechanical model for the yield strength of nanocrystalline materials. *Acta Mater.* 55: 261–271.

Lee, E.H., Mallet, R.L., and Wertheimer, T.R. (1983). Stress analysis for anisotropic hardening in finite deformation plasticity. *Trans. ASME/ J. Appl. Mech.* 50: 554–569.

Lehmann, Th. (1972). Einige Bemerkungen zu einer allgemeinen Klasse von Stofgesetzen für große elastoplastische Formänderungen. *Ing. Arch.* 41: 297–310.

Loret, B. (1983). On the effects of plastic rotation in the finite deformation of anisotropic elastoplastic materials. *Mech. Mater.* 2: 287–304.

Meyers, M.A., Mishra, A., and Benson, D.J. (2006). Mechanical properties of nanocrystalline materials. *Prog. Mater Sci.* 51: 427–556.

Nagtegaal, J.C. and De Jong, J.E. (1982). Some aspects of non-isotropic work hardening in finite strain plasticity. In: *Proceedings of the Workshop on Plasticity of Metals at Finite Strain. Theory, Experiment and Computation*, Stanford University, 1981 (ed. E.H. Lee and R. Mallet), 65–102. Stanford, Troy: Published by the Division of Applied Mechanics, Stanford University and Department of Mech. Engn. Aeronautical Engn. And Mechanics, R.P.I., Troy.

Nowak, Z. and Pęcherski, R.B. (2002). Plastic strain in metals by shear banding. II. Numerical identification and verification of plastic flow law. *Arch. Mech.* 54: 621–634.

Nowak, Z. and Stachurski, A. (2001). Nonlinear regression problem of material functions identification for porous media plastic flow. *Eng. Trans.* 49: 637–661.

Nowak, Z. and Stachurski, A. (2002). Global optimization in material functions identification for voided media plastic flow. *Comput. Assisted Mech. Eng. Sci.* 29: 205–221.

Nowak, Z., Perzyna, P., and Pęcherski, R.B. (2007). Description of viscoplastic flow accounting for shear banding. *Arch. Metall. Mater.* 52: 217–222.

Oliferuk, W., Korbel, A., and Grabski, M.W. (1966). Mode of deformation and the rate of energy storage during uniaxial tensile deformation of austenitic steel. *Mat. Sci. Eng.* A220: 123–128.

Onat, E.T. (1984). Shear flow of kinematically hardening rigid-plastic materials. In: *Mechanics of Material Behaviour* (ed. G.J. Dvorak and R.T. Shield), 311–324. Amsterdam, Oxford, New York, Tokyo: Elsevier.

Paul, H., Jasienski, Z., Piatkowski, A. et al. (1996). Crystallographic nature of shear bands in polycrystalline copper. *Arch. Metall.* 41: 337–353.

Paulun, J.E. and Pęcherski, R.B. (1985). Study of corotational rates for kinematic hardening in finite deformation plasticity. *Arch. Mech.* 37: 661–667.

Paulun, J.E. and Pęcherski, R.B. (1987). On the application of the plastic spin concept for the description of anisotropic hardening in finite deformation plasticity. *Int. J. Plast.* 3: 303–314.

Paulun, J.E. and Pęcherski, R.B. (1992a). On the relation for plastic spin. A new physical motivation. *ZAMM.* 72: T185–T190.

Paulun, J.E. and Pęcherski, R.B. (1992b). On the relation for plastic spin. *Arch. Appl. Mech.* 62: 376–385.

Pęcherski, R.B. (1988). The plastic spin concept and the theory of finite plastic deformations with induced anisotropy. *Arch. Mech.* 40: 807–818.

Pęcherski, R.B. (1995). Model of plastic flow accounting for the effects of shear banding and kinematic hardening. *ZAMM* 75: S203–S204.

Pęcherski, R.B. (1996). Finite deformation plasticity with strain induced anisotropy and shear banding. *J. Mater. Process. Technol.* 60: 35–44.

Pęcherski, R.B. (1998). Macroscopic effects of micro-shear banding in plasticity of metals. *Acta Mech.* 131: 203–224.

Pęcherski, R.B. and Korbel, K. (2002). Plastic strain in metals by shear banding. I. Constitutive description for simulation of metal shaping operations. *Arch. Mech.* 54: 603–620.

Perzyna, P. (1963). The constitutive equations for rate sensitive plastic materials. *Q. Appl. Math.* 20 (4): 321–332.

Perzyna, P. (1966). *Fundamental Problems in Viscoplasticity, Advances in Applied Mechanics*, vol. 9, 343–377. New York: Academic Press.

Perzyna, P. (1971). Thermodynamic theory of viscoplasticity. In: *Advances in Applied Mechanics*, vol. 11, 313–354. New York: Academic Press.

Stalony-Dobrzanski, F. and Bochniak, W. (2005). Role of shear bands in forming the texture image of deformed copper alloy. *Arch. Metall. Mater.* 50: 1089–1102.

Wang, H.-T. and Yang, W. (2004). Constitutive modelling for nanocrystalline metals based on cooperative grain boundary mechanisms. *J. Mech. Phys. Solids.* 52: 1151–1173.

Wei, Y. and Anand, L. (2007). A constitutive model for powder-processed nanocrystalline metals. *Acta Mater.* 55: 921–931.

Wei, Y., Su, L., and Anand, L. (2006). A computational study of the mechanical behavior of nanocrystalline FCC metals. *Acta Mater.* 54: 3177–3190.

9

Conclusions

9.1 Concluding Remarks

9.1.1 Shear Banding-Mediated Flow vis-à-vis Ductile Failure Analysis

The issue raised in the Preface on shear modes imminent to ductile fracture by low-stress triaxiality, which motivated study of the inelastic flow produced by shear banding, finds closure in this chapter. As mentioned in the Preface, numerous investigation results emphasise the role of multiscale shear modes in low-stress triaxiality ductile fracture processes. To underline this observation, one can use as an example the work results (Nguyen et al. 2020) that the most common result in the internal necking mode is the shrinking of the ligaments and a localisation bands formation. The shear coalescence mode occurs with the appearance of micro-shear bands dominating at low-stress triaxiality. The paper comprises a nonlocal modelling framework for predicting ductile failure based on a comprehensive study of micromechanics and physical mechanisms. The model's predictive capability transpires using different numerical simulations of complex failure patterns, e.g. the slant failure in the plane strain specimens, shear failure in 'V-notched' plates under shearing and the cup and cone fracture in the axisymmetric smooth and notched bars. The results of other cited papers appear and provide a wealth of knowledge of the subject.

The results of Bai and Wierzbicki (2010), as well as Ebnoether and Mohr (2013) and Mohr and Marcadet (2015), are particularly worth noting. The papers mentioned above represent a unique approach to analysing and formulating the ductile fracture criterion in metallic crack-free solids subjected to complex stress states of limited stress triaxiality. For instance, according to Mohr and Marcadet (2015), the resulting fracture initiation model predicts

Viscoplastic Flow in Solids Produced by Shear Banding, First Edition. Ryszard B. Pęcherski.
© 2022 John Wiley & Sons Ltd. Published 2022 by John Wiley & Sons Ltd.

the onset of fracture in high-strength steels (DP590, DP780, and TRIP780) at low triaxialities. The detailed and pedantically prepared experimental investigations contain pure shear, notched tension, and equi-biaxial tension tests. The observed formation of localisation bands at the mesoscale or micro-shear bands clusters that accompany directly ductile fracture is described with the extended Mohr–Coulomb stress-based criterion. The Hosford equivalent stress (Hosford 1972) and the normal stress acting on the plane of maximum shear appear to predict the onset of ductile fracture more accurately. The characteristic feature of the above-mentioned unique approach is a coherent transformation from the principal stress space to the space of equivalent plastic strain, stress triaxiality, and Lode angle parameter. Such a transition can provide a fracture-onset description for the case of non-proportional loading paths. Many practical engineering problems relate to a ductile fracture in thin-walled structures. In such a case, they require modelling and finite element method (FEM)-based numerical simulations with shell elements. Pack et al. (2018) studied the limits of application of shell elements defined in addition to the Hosford–Coulomb fracture initiation model vis-à-vis the convergence of the onset of the ductile fracture process concerning the mesh size. The study is supplemented with comprehensive plasticity and fracture testing program performed on 0.8 mm thick DP980 steel sheets, e.g. uniaxial tension, along with different material directions and simple shear, V-bending, and punch experiments. Also, 1 mm thick aluminium 6016-T4 sheets became the subject of investigations: identification and validation through the deep drawing numerical simulations.

In my view, the constitutive description of multilevel shear banding mechanisms meets possible applications in the analysis and the predictive numerical simulations of ductile failure processes. The potential applicability of the energy-based Burzynski limit criterion discussed in Chapter 7 and accounting for anisotropy in Pęcherski et al. (2021) and Moayyedian and Kadkhodayan (2015, 2021) seems to be promising both from the scientific and industrial applications point of view. The earlier experience applying the mesh-less approach for numerical simulations of deformation and failure processes (Postek et al. 2019, 2021a,b) opens the possibility of the analysis and numerical simulations predicting the mentioned ductile failure processes.

9.1.2 Application of Peridynamic Numerical Simulations of Shear Banding Processes

In the last 20 years, a new mesh-less computational approach has emerged that is generally called peridynamics. In particular, the frequently used mesh-less method is the state-based peridynamics – a continuum theory employing a nonlocal model to describe material properties. The main idea of state-based peridynamics is that spatial integration determines a force state on a material point (Silling et al. 2007). The concise presentation of the peridynamics applications and the related contributions are presented in the review work (Javili et al. 2019). In particular, the coupling of peridynamics and FEM is considered (Sun and Fish 2019).

The subject of the study (Postek et al. 2019) is the deformation of open-cell copper foams under dynamic compression with the use of the computational model of virtual cellular material (Pęcherski et al. 2017). The material of the skeleton is oxygen-free high conductivity (OFHC) copper. The foam skeleton is described as an elastic–plastic material with isotropic hardening. The dynamical processes of compression and crushing are simulated numerically based on the parallelized code. A similar problem is studied by Postek et al. (2022) using the elastic–viscoplastic model, e.g. Foster et al. (2010), incorporating a viscoplasticity theory within the state-based peridynamics framework. The OFHC copper sample undergoes an impact against an elastic wall. The dynamic processes of compression with different impact velocities are simulated. Two kinds of specimens are the subject of investigations: open-cell foams of the copper skeleton and a solid OFHC copper cylinder used in Taylor impact tests. The predicted results by the numerical viscoplasticity model of the Taylor impact test are subjected to the comparative study accounting for shear banding effects (Postek et al. 2021). The above discussion of the Taylor impact test shows that it is possible to confront the results of viscoplastic flow produced by shear banding with the experimental data. The initial results of the work by Grązka and Janiszewski (2012) open such a possibility. The recent discussions with Professor Jacek Janiszewski and the joint application of the wealth of his valuable data obtained from Taylor impact test experiments create the opportunity for confirming predictive power of the developed theory of viscoplastic flow produced by shear banding (Janiszewski et al. 2022).

The other applications of the theory of inelastic deformations produced by shear banding transpire from the results of Pieczyska et al. (2006) elaborated for the automotive industry's needs. In the glass-fibre-reinforced composite of polypropylene matrix subjected to the quasi-static compression tests on the hydraulic testing machine and dynamical loading using a modified split Hopkinson pressure bar (SHPB), the mechanism of shear banding in the polypropylene matrix plays a controlling role. The quasi-static compression was performed on the discs of 12 mm diameter and 3 mm thickness; specimens cut off from commercial polypropylene sheet composites GB30 and GB40, 30% and 40% of glass fibre, respectively. The applied strain rates are equal to 10^{-3}, 10^{-1}, 10 s^{-1}. The application of the SHPB system enables obtaining the strain rate value up to 10^3 s^{-1}. All the tests were performed at room temperature 295 K. The paper (Pieczyska et al. 2006) provides complete and detailed information about the preparation of specimens, experimental procedures, and data acquisition. The true stress–true strain characteristics obtained for both GB30 and GB40 composites, subjected to quasi-static tests and stress–strain curves calculated from the data acquired in the course of dynamical compression tests, make a basis for the specification of plasticity and viscoplasticity models, respectively, that account for the instantaneous shear-banding contribution function, cf. Section 5.2. The numerical simulations of the investigated compression processes confirm the predictive capacity of the models formulated. The hypothesis of the shear-banding contribution into the inelastic flow of polymers finds confirmation in the experimental observations performed by transmission optical microscopy and by scanning electron microscopy reported by Sell et al. (2002). The authors found that the primary mechanism of large plastic strains in solid polymers relates to the interplay of shear banding and crazing: 'The crazes are perpendicular to the tensile axis. They are 100 μm long on average and their thickness is too small to be resolved. The shear bands are inclined about 45^0 on the tensile axis and form a network of lines across the specimen' – p. 3865, (Sell et al. 2002).

The above discussion shows the new possibility of modelling the inelastic deformation of solid polymeric materials, particularly polypropylene or epoxy resins, when multilevel micro-shear banding comes into play. The results presented by Morelle (2015) confirm the observation about the crucial role of the micro-shear banding mechanism in the inelastic deformation of epoxy resin. Also, the vital contribution of the micromechanism typical for both glassy polymers and glassy metals in the form of shear transition zones (STZs) is discussed. It is worth mentioning that the yield criterion of Drucker – Prager type discussed in Morelle (2015) corresponds, in fact, according to the discussion in Chapter 7, to one of the earlier-formulated, energy-based Burzyński elastic limit conditions, cf. Frąś (2013).

An example of a recent study is epoxy resin composites doped with carbon nanoparticles (Nowak et al. 2021). The preliminary results of quasi-static axial compression tests of the resin cylindrical specimens with a dimension ratio are 1.5 : 1 (height h : diameter d), and the axial tensile tests, flat samples cut from lamellas with a thickness of 1 mm become the subject of the analysis. Tensile and compression tests were carried out for all specimens with a strain rate of 10^{-3} s^{-1}, using the materials test systems (MTSs) 858 testing machine and the digital image correlation (DIC) numerical image-processing technique. It transpires from the surface observation that resin's deterioration process results from the development of micro-shear bands leading to cracks in samples with deformation of 4% with a sharp drop in stress. Inelastic deformation in the investigated specimens is visible as the displacement and deformation of the macromolecules segments. It resembles a viscous liquid flow while maintaining a constant deformation rate in the material, revealing viscoelastic behaviour. Based on these observations, a model describing inelastic polymers' flow is necessary to account for the appropriate yield condition, cf. Frąś (2012). It corresponds to the viscoplasticity flow equation containing the shear band contribution function.

The benefit for the reader who is going through this work is to find some valuable tips on possible applications of the model of viscoplastic flow produced by shear banding in diverse industrial branches. The above discussion shows a broad spectrum of engineering applications in which the studied multilevel micro-shear bands' mechanism is responsible for plastic or viscoplastic deformations and failure. The proper account for shear banding contribution function, instantaneous one or having cumulative volumetric character, can provide more adequate predictions of the considered approaches. The reference mentioned above to a kind of mesh-less approach – peridynamics – can guide the reader to possible applications in the numerical simulations and analysis of the plastic or viscoplastic processes leading to failure prediction and experimental verification.

References

Bai, Y. and Wierzbicki, T. (2010). Application of extended Mohr–Coulomb criterion to ductile fracture. *Int. J. Fract.* 161: 1–20.

Ebnoether, F. and Mohr, D. (2013). Predicting ductile fracture of low carbon steel sheets: stress-based versus mixed stress/strain-based Mohr–Coulomb model. *Int. J. Solids Struct.* 50: 1055–1066.

Foster, J.T., Silling, S.A., and Chen, W.W. (2010). Viscoplasticity using peridynamics. *Int. J. Numer. Methods Eng.* 81: 1242–1258.

Frąś, T. (2013). Modelling of plastic yield surface of materials accounting for initial anisotropy and strength differential effect on the basis of experimental and numerical simulation. PhD thesis. Metz & Krakow: Université de Lorraine & AGH University of Science and Technology.

Grązka, M. and Janiszewski, J. (2012). Identification of Johnson–Cook equation constants using finite element method. *Eng. Trans.* 60: 215–223.

Hosford, W.F. (1972). A generalized isotropic yield criterion. *J. Appl. Mech.* 39: 607–609.

Janiszewski, J., Postek, E., Nowak, Z., Frąś, L., and Pęcherski, R.B. (2022). Verification of viscoplastic flow produced by shear banding using Taylor impact test experiment – under preparation for publication.

Javili, A., Morasata, R., Oterkus, E., and Oterkus, S. (2019). Peridynamics review. *Math. Mech. Solids* 24: 3714–3739.

Moayyedian, F. and Kadkhodayan, M. (2015). Modified Burzynski criterion with non-associated flow rule for anisotropic asymmetric metals in plane stress problem. *Appl. Math. Mech. – Engl. Ed.* 36: 303–318.

Moayyedian, F. and Kadkhodayan, M. (2021). Modified Burzynski criterion along with AFR and non-AFR for asymmetric anisotropic materials. *Arch. Civil Mech. Eng.* 64: 2–18.

Mohr, D. and Marcadet, S.J. (2015). Micromechanically - motivated phenomenological Hosford–Coulomb model for predicting ductile fracture initiation at low stress triaxialities. *Int. J. Solids Struct.* 67–68: 40–55.

Morelle, X. (2015). *Mechanical Characterization and Physics-Based Modeling of Highly-Crosslinked Epoxy Resin* (Prom. T. Pardoen and C. Bailly). UCL Louvain.

Nguyen, V.-D., Pardoen, T., and Noels, L. (2020). A nonlocal approach of ductile failure incorporating void growth, internal necking, and shear dominated coalescence mechanisms. *J. Mech. Phys. Solids* 137: 103891.

Nowak, Z., Giersig, M., and Pęcherski, R.B. (2021). Experimental investigation of the quasi-static deformation of an epoxy resin. In: *Proceedings of the XIII Scientific Conference Integrated Studies of the Foundations of Plastic Deformation of Metals*, 23–26 November, Łańcut Castle (ed. B. Pawłowska and R.E. Śliwa). Poland: Publishing House. Rzeszów University of Technology.

Pack, K., Tancogne-Dejean, T., Gorji, M.B., and Mohr, D. (2018). Hosford–Coulomb ductile failure model for shell elements: Experimental identification and validation for DP980 steel and aluminum 6016-T4. *Int. J. Solids Struct.* 151: 214–232.

Pęcherski, R.B., Nowak, M., and Nowak, Z. (2017). Virtual metallic foams, applications for dynamic crushing analysis. *Int. J. Multiscale Comput. Eng.* 15: 431–442.

Pęcherski, R.B., Rusinek, A., Frąś, T., Nowak, M. and Nowak, Z. (2021). Energy-based yield condition for orthotropic materials exhibiting asymmetry of elastic range. *Arch. Metall. Mater.* 65: 771–778.

Pieczyska, E.A., Pęcherski, R.B., Gadaj, S.P., Nowacki, W.K., Nowak, Z., Matyjewski, M. (2006). Experimental and theoretical investigations of glass-fibre reinforced composite subjected to uniaxial compression for a wide spectrum of strain rates. *Arch. Mech.* 58: 273–291.

Postek, E., Pęcherski, R.B., and Nowak, Z. (2019). Peridynamic simulation of crushing processes in copper open-cell foam. *Arch. Metall. Mater.* 64: 1603–1610.

Postek, E., Nowak, Z., and Pęcherski, R.B. (2022). Viscoplastic flow of functional cellular materials with use of peridynamics. *Meccanica*. 57: 905–922.

Postek, E., Nowak, Z., and Pęcherski, R. (2021). Viscoplastic material with shear banding effects. *16th US National Congress on Computational Mechanics*, 25–29 July 2021. Chicago. Illinois (A virtual event), submitted for publication in Composite Structures.

Sell, C.G., Hiver, J.M., and Darhoun, A. (2002). Experimental characterization of deformation damage in solid polymers under tension, and its interrelation with necking. *Int. J. Solids Struct.* 39: 3857–3872.

Silling, S.A., Epton, E., Weckner, O. et al. (2007). Peridynamic states and constitutive modelling. *J. Elast.* 88: 151–184.

Sun, W. and Fish, J. (2019). Superposition-based coupling of peridynamics and finite element method. *Comput. Mech.* 64: 231–248.

Subject Index

Viscoplastic Flow in Solids Produced by Shear Banding, First Edition. Ryszard B. Pęcherski.
© 2022 John Wiley & Sons Ltd. Published 2022 by John Wiley & Sons Ltd.

Name Index

Viscoplastic Flow in Solids Produced by Shear Banding, First Edition. Ryszard B. Pęcherski.
© 2022 John Wiley & Sons Ltd. Published 2022 by John Wiley & Sons Ltd.